Novel Pharmacological Inhibitors for Bacterial Protein Toxins

Special Issue Editor
Holger Barth

MDPI

Special Issue Editor
Holger Barth
Universitätsklinikum Ulm
Germany

Editorial Office
MDPI AG
St. Alban-Anlage 66
Basel, Switzerland

This edition is a reprint of the Special Issue published online in the open access journal *Toxins* (ISSN 2072-6651) from 2016–2017 (available at: http://www.mdpi.com/journal/toxins/special_issues/pharma-inhibitor).

For citation purposes, cite each article independently as indicated on the article page online and as indicated below:

Author 1; Author 2; Author 3 etc. Article title. *Journal Name.* **Year**. Article number/page range.

ISBN 978-3-03842-430-7 (Pbk)
ISBN 978-3-03842-431-4 (PDF)

Table of Contents

About the Guest Editor

Holger Barth studied Biology with a focus on Microbiology at the University of Darmstadt (Germany) and received his PhD in 1994 with a dissertation prepared at the German Cancer Research Center in Heidelberg (Germany). From 1995 to 2004, he was first a group leader in the laboratory of Professor Klaus Aktories and later Assistant Professor at the Institute for Experimental and Clinical Pharmacology and Toxicology of the University of Freiburg (Germany), where he started his work on bacterial protein toxins. In 2002, he received his habilitation and in 2005 he became a registered German and EU board toxicologist. Since 2004, he has been Professor for Pharmacology and Toxicology at the University of Ulm (Germany).

Holger Barth is Editor of the textbook Toxikologie and an editorial board member of various journals in the field of Toxinology, Toxicology and Pharmacology. He is the chairman of the section "Biological Toxins" within the German Society of Toxicology (GT) and currently he serves as President of the GT. In 2017, Holger Barth also became the President of the German Society of Experimental and Clinical Pharmacology and Toxicology (DGPT). Holger Barth´s research interests focus on the cellular uptake of bacterial protein toxins, in particular ADP-ribosylating toxins. His group discovered the role of host cell chaperones in intracellular membrane translocation of such toxins and works on novel pharmacological strategies to inhibit toxin uptake into mammalian cells. Professor Barth's laboratory exploits the transport subunits of various bacterial toxins to deliver pharmacologically active molecules, e.g., therapeutic enzymes and peptides, into the cytosol of human target cells, in particular monocytes/macrophages, to modulate cell functions in the context of traumatic injury, inflammation and cancer.

Preface to "Novel Pharmacological Inhibitors for Bacterial Protein Toxins"

Many medically relevant bacteria cause severe human and animal diseases because they produce and release protein toxins that target mammalian cells. Because the toxin-induced cell damage is the reason for the clinical symptoms, the targeted pharmacological inhibition of the cytotoxic mode of action of bacterial toxins should prevent or cure the respective toxin-associated disease. Toxin inhibitors might be beneficial when the toxin acts in the absence of the producing bacteria (e.g., food poisoning), but also in combination with antibiotics in infectious diseases when the toxin-producing bacteria are present. The focus of this Special Issue of Toxins is on the development and characterization of novel inhibitors against bacterial toxins, e.g., toxin neutralizing antibodies, peptides or small compounds, as well as toxin pore blockers, which interfere with bacterial toxins and thereby protect cells from intoxication.

Holger Barth
Guest Editor

Editorial

An Introduction to the *Toxins* Special Issue on "Novel Pharmacological Inhibitors for Bacterial Protein Toxins"

Holger Barth

Institute of Pharmacology and Toxicology, University of Ulm Medical Center, Albert-Einstein-Allee 11, 89081 Ulm, Germany; holger.barth@uni-ulm.de

Academic Editor: Vernon L. Tesh
Received: 9 May 2017; Accepted: 10 May 2017; Published: 11 May 2017

Bacterial AB-type protein toxins that consist of an enzymatically active subunit (A) and a binding/transport subunit (B), are among the most toxic substances and represent the causative agents for a variety of severe human and animal diseases, such as in the context of infections, post-traumatic complications or food poisoning. Moreover, some AB-type toxins can be misused as biological warfare agents and in the context of bio-terrorism activities. Therefore, novel pharmacological inhibitors against such toxins are urgently needed. The remarkable toxicity of AB-type toxins is due to their unique modular structure and mode of action. Their B subunit very efficiently mediates the transport of the A subunit into the cytosol of mammalian target cells where the A subunit modifies its specific substrate molecules resulting in cell damage. Thus, such toxins are highly potent as well as being very substrate-specific enzymes that act inside cells, thereby causing the characteristic diseases associated with the individual toxins and toxin-producing bacteria. In past years, significant progress has been made in understanding the molecular mechanisms underlying the cellular uptake of bacterial AB-type toxins. This process involves receptor-binding, receptor-mediated endocytosis, intracellular transport in vesicles and finally the translocation of the enzyme subunit across cellular membranes into the cytosol. It was discovered that some toxins form pores in endosomal membranes and exploit host cell factors, such as chaperones, for the transport of their A subunit across intracellular membranes into the cytosol. This translocation can occur either from acidified endosomes, as first reported for Diphtheria toxin, or later in the cell from the endoplasmic reticulum, as described originally for Cholera toxin and other AB_5 toxins, including the Pertussis toxin and Shiga toxin.

Increasing knowledge about the cellular uptake has allowed for the development or screening of pharmacological compounds that inhibit the individual steps of toxin uptake and protect cells from intoxication, which is described in various chapters of this book. For all these inhibitors, the final consequence is the same and independent of which state of toxin uptake or transport in the cell they prevent. As long as the A subunit does not reach the cytosol, the cytotoxic effects do not occur if the substrate for the respective toxin remains in the cytosol. In conclusion, compounds that act on the level of the toxins rather than on the toxin-producing bacteria should have attractive therapeutic perspectives, including the following:

- They can serve as anti-toxins when only the toxins, but not the toxin-producing bacteria, are taken up and enter the body, such as botulinum neurotoxin (food poisoning).
- During infections, they would inhibit the mode of action of the toxins, which are already released by the bacteria and therefore could be combined with antibiotics. This should be of particular interest when the infection is caused by toxin-producing bacteria which are (multi-)resistant towards the classical antibiotics.

- In contrast to the toxin-neutralizing antibodies (antisera), they act on toxins that are already internalized into their target cells, as long as the A subunit did not reach the cytosol. Neutralizing antibodies are still not available for each toxin.
- Some compounds inhibit cellular uptake of toxin families (e.g., Adenosine diphosphate (ADP)-ribosyltransferases, pore-forming toxins) and therefore could serve as potential therapeutics against all toxins that are members of such toxin families.

As it becomes evident from the various chapters of this book, many different types of inhibitors have been identified and their mode of action has been analyzed in detail on the biochemical and cellular levels. Most of the presented research includes in vitro data with cell-based test systems, animals or human organoids. Thus, further efforts are required to transfer this promising novel knowledge into therapeutic strategies.

The first chapter of this book from Kirsten Sandvig's group provides an excellent overview on the uptake and intracellular transport of bacterial toxins with the focus on the Shiga toxins. These AB_5 toxins are produced by various bacteria and associated with severe human diseases, such as the hemolytic-uremic syndrome (HUS). In this chapter, a variety of pharmacological inhibitors are reviewed, which inhibit each individual step during the cell entry of Shiga toxins from receptor-binding to translocation of the A subunit from the endoplasmic reticulum into the host cell cytosol [1].

In the second chapter, the Cheng group shows that treatment with several probiotic micro-organisms, including Saccharomyces and Lactobacillus strains, but not with a non-probiotic strain of *Escherichia coli*, inhibits the internalization of *Clostridium botulinum* neurotoxin serotype A (BoNT/A) into human epithelial cells (CaCo-2) in vitro. The probiotic strains do not bind BoNT A or cause its degradation, although the findings suggest that there is some competition between the strains and BoNT/A for binding to the cell surface [2].

The third chapter from Bruce McClane's group (Li et al.) suggests sialidase inhibitors as potential novel therapeutics to treat/prevent intestinal infections caused by *Clostridium perfringens*. *C. perfringens* causes histotoxic infections with traumatic gas gangrene and myonecrosis as well as intestinal infections of humans and animals. *C. perfringens* produces up to three sialidases that are involved in the histotoxic effects. The sialidases may also contribute to the intestinal infection, because they upregulate the production of some toxins, their activity and their binding to the surface of target cells. Moreover, there is some evidence that the sialidase, NanI, might contribute to the intestinal colonization by *C. perfringens*, because strains that cause acute food poisoning lack the NanI gene while strains that cause chronic intestinal infections carry this gene [3].

The fourth chapter from the Genth group [4] and fifth chapter from the Barth laboratory [5] both describe compounds that inhibit the cellular uptake of single chain AB-type toxins, namely the *Clostridium sordellii* Lethal toxin and Diphtheria toxin, respectively. After their receptor-mediated endocytosis, both toxins deliver their A subunit from acidic endosomes into the cytosol. Schelle and co-workers identified the p38$_{alpha/beta}$ MAP kinase inhibitor SB203580 as an inhibitor against the Lethal toxin and found that this compound might inhibit the toxin uptake rather than the enzymatic reaction in the cell, because it inhibits the Lethal toxin but not the related TcdB from *C. difficile*, which has the same enzymatic mode of action. Schnell et al. identified EGA, a semicarbazone compound, as a potent inhibitor against Diphtheria toxin and showed that EGA prevents the pH-dependent transport of the A subunit of this toxin across cell membranes.

The last two chapters focus on compounds that block the heptameric trans-membrane pores formed by the B subunits of binary toxins from *Bacillus anthracis* in addition to the C2 and iota toxins from *Clostridium botulinum* and *C. perfringens*, respectively, in endosomal membranes under acidic conditions. These pores serve as translocation channels for the A subunits of these toxins, which unfold to translocate through the pores across endosomal membranes into the host cell cytosol. Pharmacological pore-blockers prevent this step and protect cells from intoxication by anthrax and clostridial binary toxins. The chapter from the Nestorovich group [6] describes pore-blockers, which are directed against the PA63 pore of the anthrax toxins and are based on cyclic dendrimers. These

Toxins **2017**, *9*, 160

compounds efficiently block the pores formed by the PA63 component of anthrax toxin in artificial lipid bilayer membranes, which enables single molecule analysis in biophysical in vitro approaches. The Benz and Barth groups [7] characterized compounds derived from chloroquine, which inhibit the heptameric trans-membrane pores of the binary clostridial C2 and iota toxins in lipid bilayers in vitro, and demonstrated the protective effects of the most efficient pore-blockers in cell-based experiments.

References

1. Kavaliauskiene, S.; Dyve Lingelem, A.B.; Skotland, T.; Sandvig, K. Protection against Shiga Toxins. *Toxins* **2017**, *9*, 44. [CrossRef] [PubMed]
2. Lam, T.I.; Tam, C.C.; Stanker, L.H.; Cheng, L.W. Probiotic Microorganisms Inhibit Epithelial Cell Internalization of Botulinum Neurotoxin Serotype A. *Toxins* **2016**, *8*, 377. [CrossRef] [PubMed]
3. Li, J.; Uzal, F.A.; McClane, B.A. *Clostridium perfringens* Sialidases: Potential Contributors to Intestinal Pathogenesis and Therapeutic Targets. *Toxins* **2016**, *8*, 341. [CrossRef] [PubMed]
4. Schelle, I.; Bruening, J.; Buetepage, M.; Genth, H. Role of $p38_{alpha/beta}$ MAP Kinase in Cell Susceptibility to *Clostridium sordellii* Lethal Toxin and *Clostridium difficile* Toxin B. *Toxins* **2017**, *9*, 2. [CrossRef] [PubMed]
5. Schnell, L.; Mittler, A.-K.; Mattarei, A.; Tehran, D.A.; Montecucco, C.; Barth, H. Semicarbazone EGA Inhibits Uptake of Diphtheria Toxin into Human Cells and Protects Cells from Intoxication. *Toxins* **2016**, *8*, 221. [CrossRef] [PubMed]
6. Yamini, G.; Kalu, N.; Nestorovich, E.M. Impact of Dendrimer Terminal Group Chemistry on Blockage of the Anthrax Toxin Channel: A Single Molecule Study. *Toxins* **2016**, *8*, 337. [CrossRef] [PubMed]
7. Kronhardt, A.; Beitzinger, C.; Barth, H.; Benz, R. Chloroquine Analog Interaction with C2- and Iota-Toxin in Vitro and in Living Cells. *Toxins* **2016**, *8*, 237. [CrossRef] [PubMed]

toxins

MDPI

Review
Protection against Shiga Toxins

Simona Kavaliauskiene [1,2], Anne Berit Dyve Lingelem [1,2], Tore Skotland [1,2] and Kirsten Sandvig [1,2,3,*

1   Department of Molecular Cell Biology, Institute for Cancer Research, Oslo University Hospital, N-0379 Oslo, Norway; simona.kavaliauskiene@rr-research.no (S.K.); Anne.Berit.Dyve@rr-research.no (A.B.D.L.). Tore.Skotland@rr-research.no (T.S.)
2   Center for Cancer Biomedicine, Faculty of Medicine, Oslo University Hospital, N-0379 Oslo, Norway
3   Department of Biosciences, University of Oslo, N-0316 Oslo, Norway
*   Correspondence: kirsten.sandvig@ibv.uio.no; Tel.: +47-22-78-18-28

Academic Editor: Holger Barth
Received: 29 December 2016; Accepted: 19 January 2017; Published: 3 February 2017

Abstract: Shiga toxins consist of an A-moiety and five B-moieties able to bind the neutral glycosphingolipid globotriaosylceramide (Gb3) on the cell surface. To intoxicate cells efficiently, the toxin A-moiety has to be cleaved by furin and transported retrogradely to the Golgi apparatus and to the endoplasmic reticulum. The enzymatically active part of the A-moiety is then translocated to the cytosol, where it inhibits protein synthesis and in some cell types induces apoptosis. Protection of cells can be provided either by inhibiting binding of the toxin to cells or by interfering with any of the subsequent steps required for its toxic effect. In this article we provide a brief overview of the interaction of Shiga toxins with cells, describe some compounds and conditions found to protect cells against Shiga toxins, and discuss whether they might also provide protection in animals and humans.

Keywords: Shiga toxin; Stx1; Stx2; hemolytic uremic syndrome; inhibitors; chloroquine; fluorodeoxyglucose; Mn^{2+}

1. Introduction

Shiga toxins (Stxs) comprise a family of related bacterial protein toxins that are similar in structure and mechanism of action, but are produced by different types of bacteria. Shiga toxin is secreted by *Shigella dysenteriae*, whereas Shiga-like toxin 1 (Stx1) and Shiga-like toxin 2 (Stx2) are produced by certain strains of *Escherichia coli* (Shiga toxin-producing *E. coli* (STEC)) and some other bacteria [1]. Prototypic Stx1 (Stx1a) differs from Shiga toxin only in one amino acid residue in the catalytic A-moiety of the toxin, whereas Stx2 shares only ~60% sequence similarity with Shiga toxin and defines an immunologically distinct subgroup comprised of at least seven subtypes of Stx2 [2]. Stx2 is more lethal than Stx1 in animal models [3,4] and is thought to be the main cause of life-threatening infections in humans. Some STEC produce only one toxin type, either Stx1 or Stx2, while others express a combination of both types and different subtypes [5]. For simplicity, we will use the abbreviation Stx to refer to the whole family of Shiga toxins when discussing general facts about the toxin and/or where the exact type or variant is not known.

Infection with enterohemorrhagic STEC may cause hemorrhagic colitis, hemolytic uremic syndrome (HUS), and death [6]. There is no approved treatment of STEC-induced HUS, and the use of antibiotics may worsen the disease by increasing toxin formation and release by the bacteria [7]. In general, HUS occurs in 5%–15% of cases with STEC infection, with children having the highest risk [8], although the large outbreak with a Stx2a-producing enteroaggregative STEC strain in Northern Europe in 2011 demonstrated that there are bacterium-toxin combinations that can be as dangerous to adults as to children [9]. HUS will most often occur 5–13 days after the onset of diarrhea, with a mortality of 3%–5% [10,11]. In addition to direct renal damage, neurological complications may also occur in HUS patients and are important determinants of severity of the condition and mortality rate [12–15]. Neurological symptoms may be caused by fatigue, cerebral microvascular thrombi, ischemia-hypoxia, or the direct neuronal effects of Stxs [12,14,16].

One of the first specific therapeutic approaches against infections with Stxs was the idea of sequestering the toxin once it is released in the gut. In this regard, a novel agent composed of silicon dioxide particles covalently linked to the trisaccharide moiety of the globotriaosylceramide molecule that mediates Stx binding (Synsorb® Pk, Synsorb Biotech) was developed. However, although Synsorb® Pk was shown to bind and neutralize Stx1 (and Stx2, but less efficiently) in vitro [17], it failed to improve the clinical course of diarrhea-associated HUS in pediatric patients when tested in a randomized clinical trial [18]. The main drawback of neutralization of Stxs in the intestine for the prevention of HUS is that only trace amounts of the toxin reaching circulation are sufficient to induce HUS, and thus a more systemic treatment is required. Taking this into account, analogues of the globotriaosylceramide (Gb3) receptor and Stx antibodies for systematic administration have been developed and proven promising in in vivo models [19–21]. In addition, human serum amyloid component P (HuSAP) has been found to neutralize Stx2, but not Stx1, in vitro [22], and to protect mice against a lethal dose of Stx2 [23]. Moreover, eculizumab, an antibody directed against the complement protein C5, was used in patients with HUS during the outbreak in Northern Europe in 2011 [24] in order to counteract the activation of complement by the toxin [25]. These novel strategies based on direct neutralization of Stx in the intestine and/or circulation and the inhibition of complement have been well described in a recent review by Melton-Celsa and O'Brien [26] and thus are not further discussed here. In this review we will first provide a short overview of the toxin structure, toxin binding to the glycosphingolipid Gb3, and the intracellular transport, before we focus on the potential therapeutic agents for treatment of STEC infections and HUS that target specific cellular functions and protect cells against Stx by inhibiting toxin binding and/or intracellular trafficking.

1.1. Stx Structure

Stxs belong to the AB_5 class of protein toxins and consist of an A-moiety (~32 kDa), which is non-covalently attached to a homo-pentameric B-moiety (7.7 kDa per monomer) (Figure 1) [27,28]. Nearly all Stxs bind exclusively to the globotriaosylceramide Gb3 [29–31] with the exception of one Stx2 subtype, Stx2e, which has been shown to bind to Gb4 [32]. Each B subunit harbors three Gb3 binding sites [33], making the toxin capable of binding up to 15 Gb3 molecules on the cell surface (Figure 1C). However, not all binding sites have equal affinity for the carbohydrates of Gb3 [34,35] and, therefore, not all sites might be required for binding to the cell surface, but might rather mediate additional recognition. The B-moiety alone is not toxic to cells (with the exception of B cells, where it may induce apoptosis [36]) and functions as a delivery tool for the enzymatically active A-moiety. It is still not clear why Stx2 is more lethal to humans than Stx1, but crystallographic studies and investigations of deletion mutants reveal important differences when it comes to the role of the C-terminal end of the A-subunit for retrograde transport and complex stability [28,37,38].

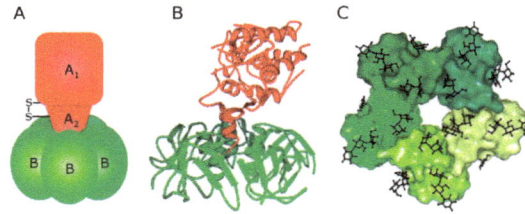

Figure 1. The schematic and structural models of Shiga toxins (Stxs). (**A**) Stxs consist of two non-covalently linked moieties: an A-moiety of ~32 kDa (shown in red), and a B-moiety (shown in green), comprised of five 7.7 kDa B-chains [27,28]. During intracellular toxin transport, the A-moiety is cleaved by the protease furin [39] into two fragments: an enzymatically active A_1 fragment (~27 kDa) and a carboxyl terminal A_2 fragment, which remain linked by a disulfide bond until arrival to the endoplasmic reticulum (ER) [40]. (**B**) The structure of the holotoxin as determined by *X*-ray crystallography [28] (PDB ID:1DM0); (**C**) The receptor-binding surface of the B-pentamer based on the structure of Stx1 complexed with the Gb3 analogue MCO-PK (methoxycarbonyloctyl glycoside of P^k trisaccharide) [33] (PDB ID:1BOS); the sugar moieties of MCO-PK are shown in black. Structure images were prepared using PDB ProteinWorkshop 4.2.

1.2. Gb3 and Its Interaction with Stx

Globotriaosylceramide (Gal-α1→4Gal-β1→4Glc-β1→Cer, Gb3; Figure 2) is a glycosphingolipid expressed on the surface of certain cell types. Gb3 is formed by the addition of one galactose residue to lactosylceramide (LacCer), which is a common precursor for different classes of glycosphingolipids, and the reaction is catalyzed by Gb3 synthase (lactosylceramide α-1,4-galactosyltransferase). Gb3 is the first glycosphingolipid in the globo-series and thus serves as a precursor for the synthesis of more complex globo-series glycosphingolipids, such as globotetraosylceramide (Gb4). Gb4 is formed after addition of *N*-acetylgalactosamine (GalNAc) to the terminal galactose of Gb3. Although Gb3 is the primary receptor for all Stxs, it has been suggested that Gb4 might facilitate Stx2 binding to colon epithelium cells, which normally have no Gb3 or very low levels of Gb3 [41].

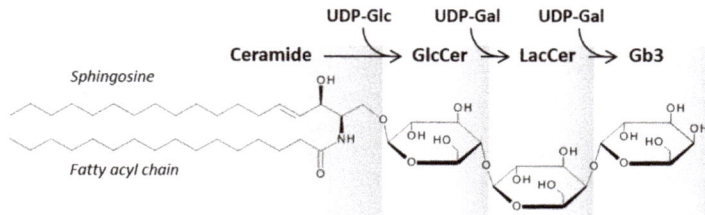

Figure 2. Chemical structure and biosynthesis of the Stx receptor globotriaosylceramide (Gb3). Sphingosine most often contains 18 carbon atoms, whereas the fatty acyl chain of ceramide varies both in length and saturation (here shown as C16:0). Gb3 is synthesized from LacCer by the addition of one galactose, and the reaction is catalyzed by Gb3 synthase (lactosylceramide α-1,4-galactosyltransferase). The sugar chain for Gb3 is: Gal-α1→4Gal-β1→4Glc-β1→Ceramide.

The sphingosine chain in the ceramide part of Gb3 is relatively invariable (most often it is monounsaturated with 18 carbon atoms, i.e., d18:1), but the *N*-amidated fatty acyl chain varies both in length (most common are 16–24 carbon atoms) and saturation resulting in multiple Gb3 species present in cells. Importantly, the receptor function of Gb3 has been shown to depend on its species composition [42–45], which in turn depends on cell type [46] and growth conditions [47], and might change in response to certain treatments, like exposure to butyric acid and cytokines [43,48–52]. It has been suggested that the production of butyric acid by the bacterial flora in the intestine may affect the expression and composition of Gb3 in the target cells and in turn lead to different susceptibility to the toxin between individuals [53]. In addition, the turnover time of Gb3 in the cells depends on its species, with longer fatty acyl chain-containing species having a longer half-life than the species with short fatty acyl chain [54]. Thus, inhibition of Gb3 synthesis will primarily lead to changes in Gb3 species composition in cells [43,50]. Studies based on artificial systems, where Gb3 was immobilized on thin layer chromatography (TLC) or ELISA plates, have shown that Stx1 and Stx2 have different binding preferences for different Gb3 species [44,55], although a mixture of various Gb3 species was required for the highest binding affinity [45]. Stx1 binding to Gb3 has also been shown to depend on cholesterol levels in the membrane [56,57]. Stx2 has been shown to be more potent in mice [3] and is more often associated with disease in humans [58], although the binding affinity of Stx2 to Gb3 is lower than that of Stx1, when measured using Gb3 adsorbed on a microtiter plate [59], and Stx2 is less toxic to Vero cells than Stx1 [60]. The different pathology observed for Stx1 and Stx2 might be caused by differences in receptor binding and thus differential targeting to susceptible tissues [61–63], as well as differences in intracellular transport of the toxins [37,64–66].

In the human body, the expression of Gb3 is restricted to only certain tissues. Normally, the highest Gb3 content is found in the microvascular glomeruli and proximal tubule cells of the kidney, consistent with the renal pathology of HUS [63,67–69]. Gb3 is also found in microvascular endothelial cells, and during infection with STEC, main Stx-target sites are the vascular endothelium of the colon [70,71] and the central nervous system [16,72]. Moreover, Gb3 is expressed in platelets [73,74], and in the carbohydrate defined P histo-blood group system, Gb3 constitutes the rare P^k antigen present on erythrocytes [75]. In the immune system, Gb3 represents a lymphocyte differentiation antigen, termed CD77, which is expressed in a subset of germinal center B lymphocytes [76]. It should also be noted that Gb3 expression is frequently increased in cancer cells [77]. However, the physiological role of Gb3 is still unclear, and it is not known why Gb3 expression is restricted to certain tissues. In vivo studies of Gb3 synthase knock-out mice, which displayed a total loss of Gb3 and other globo-series glycosphingolipids, showed no changes in birth-rates and no apparent abnormalities over a year of nurturing, with the exception of total loss of sensitivity to Stx1 and Stx2 as compared to wild-type mice [78].

1.3. Intracellular Transport of Stx

Upon binding to cells, the toxin has been found to activate a number of tyrosine kinases, including Syk [79], and the Src kinases Yes [80] and Lyn [81], as well as the serine/threonine protein kinase Cδ (PKCδ) [82] and the mitogen-activated protein kinase (MAPK) p38α [83]. Although the exact mechanism of how Stx mediates these signaling events is not yet fully understood, a recent study from our group have shown that the activation of Syk depends on the multivalent cross-linking of Gb3 at the plasma membrane, which in turn leads to an increase in cytosolic calcium levels and phosphorylation of Syk [84]. In addition, StxB binding to the cells has been shown to induce the release of cytoplasmic phospholipase A2 (cPLA2) from a cPLA2-annexin A2 complex, thereby facilitating Golgi transport, which has been found to be dependent on cPLA2 [85]. Furthermore, binding of the Stx B-moiety has been reported to stimulate remodeling of cytoskeleton components, such as actin, ezrin and dynein [86–88]. Thus, Stx is able to induce cell signaling and to modulate various cellular components to favor its uptake and intracellular transport.

Receptor-bound Stx becomes internalized by different endocytic mechanisms, including both clathrin- and dynamin-dependent and independent pathways [1]. After internalization, the toxin is transported from early/recycling endosomes directly to the Golgi apparatus [89] and then further to the endoplasmic reticulum (ER) [90–93]. During the transport, the A-moiety is cleaved by the protease furin, leaving two fragments, A_1 and A_2 (Figure 1), which remain linked to each other by a disulfide bridge [39]. Cleavage of Stx is optimal at low pH [94], indicating that it can occur early in the transport pathway. However, cells that lack furin can also cleave Stx, but less efficiently and at a later stage of the transport [39,95]. In the ER, the disulfide bond between the A_1 and A_2 subunits is reduced and the

A_1-subunit is released from the toxin. Finally, the A_1-subunit is translocated across the ER membrane and inhibits protein synthesis by cleaving one adenine residue from the 28S RNA of the 60S ribosomal subunit [1]. However, the action of Stx in the cells is not limited to the inhibition of protein synthesis, and other cellular responses, such as cytokine expression and apoptosis, are triggered by the toxin (for review see [96,97]). Thus, efficient intracellular Stx transport depends on various cellular proteins and kinases, ER chaperones and other factors (for review see [1]). Drugs that affect any of these factors might interfere with proper intracellular transport of Stx and protect cells against the cytotoxic action of the toxin. In the next sections we will give an overview of compounds shown to protect cells against Stx, and we will discuss mechanisms responsible for the protection against the toxin, as well as the potential applicability of these drugs in vivo. An overview of the compounds discussed here and how they might act against Stx is given in Table 1. The intracellular transport of Stx and which steps of the transport are affected by the different compounds are shown in Figure 3.

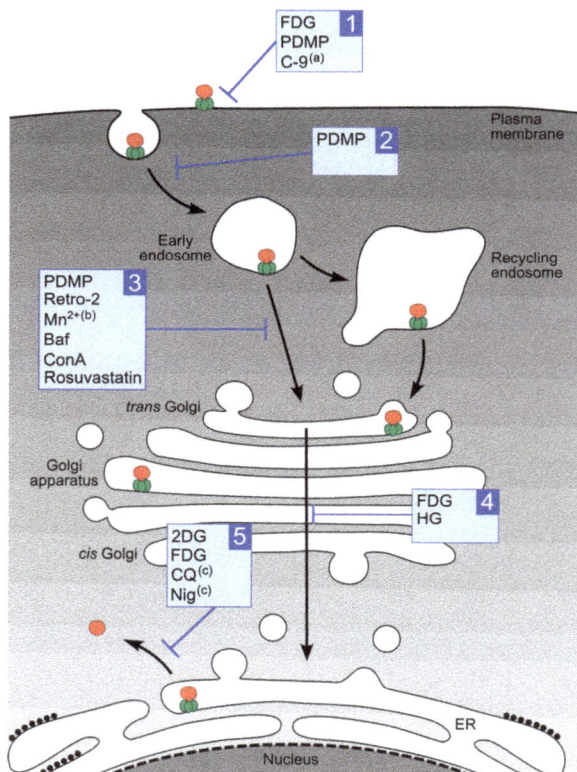

Figure 3. Stx uptake and intracellular transport, and the steps affected by different compounds. Stx binds to Gb3 on the cell surface and is taken up by various endocytic mechanisms. Following endocytosis, the toxin is transported through early endosomes and recycling endosomes and to the Golgi apparatus. From the Golgi, Stx is transported retrogradely to the ER, where its catalytically active A_1-subunit is released and translocated into the cytosol. The different compounds discussed in this review are shown with their suggested action on different steps of Stx intoxication: 1—Stx binding; 2—Stx endocytosis; 3—Stx sorting to the Golgi; 4—Stx transport via Golgi to ER; 5—release and translocation of StxA$_1$; (a) predicted effect for Stx2; (b) no effect for Stx2; (c) predicted effect for Shiga toxin.

Table 1. Compounds that protect cells against Stx.

Compound	Cellular Action	Targeted Step of Stx Intoxication	Cell Lines Tested	In Vivo Studies	Reference(s)
CQ	Elevation of pH in acidified organelles	Translocation of A_1-moiety to cytosol (predicted)	HEp-2	-	[98]
Baf	V-ATPase inhibitor	Transport to the Golgi	HEp-2	-	[98]
ConA	V-ATPase inhibitor	Transport to the Golgi	HEp-2	-	[98]
Nig	Ionophore that exchanges H^+ for monovalent cations	Not determined; later than Golgi	HEp-2	-	[98]
2DG	Inhibition of glycolysis and protein N-glycosylation; Ca^{2+} release from the endoplasmic reticulum (ER); Inhibition of Gb3 synthesis	Release of A_1-moiety	HEp-2, HT-29, SW480, HeLa	-	[99]
FDG	Inhibition of glycolysis and protein N-glycosylation; Ca^{2+} release from the ER; Inhibition of GlcCer synthesis	Binding; Transport from Golgi to ER; Release of A_1-moiety	HEp-2, HT-29, MCF-7, HBMEC	-	[48]
Retro-2 substances	Relocalization of Syntaxins 5 and 6	Transport from endosomes to the Golgi	A459, HeLa	Reduction in mortality rate from 70% to 40% in mice infected with *E. coli* O104:H4	[100,101]
Mn^{2+}	Induction of GPP130 oligomerization and its sorting to lysosomes for degradation	Transport from endosomes to the Golgi (no effect on Stx2 transport)	HeLa, Vero	Protection against lethal doses of Stx1 in BALB/c mice; No protection against either Stx1 or Stx2 in CD-1 mice	[65,102–105]
PDMP	Inhibition of GlcCer synthesis	Binding and endocytosis; Transport from endosomes to the Golgi	HEp-2	-	[43]
C-9	Inhibition of GlcCer synthesis	Not investigated	Human renal tubular epithelial cells, Human glomerular endothelial cells	50% reduction in mortality rate in rats injected with supernatant from *E. coli* expressing Stx2	[106–108]
HG	Ether lipid precursor	Transport from Golgi to ER	HEp-2, HMEC-1, HBMEC	-	[109]
Rosuvastatin	Inhibition of cholesterol biosynthesis and protein prenylation	Transport to the Golgi	ACHN	-	[110]
Furin inhibitors	Inhibition of furin	Proteolytic cleavage of the A-moiety	HEp-2	-	[111,112]

2. Compounds that Protect Cells against Stx

2.1. Chloroquine

Chloroquine (CQ; N'-(7-chloroquinolin-4-yl)-N,N-dietyl-pentane-1,4-diamine) is a weak base that in its unprotonated form easily can diffuse across membranes and accumulate in acidic compartments of the cell. There, CQ becomes protonated and trapped, leading to an elevated pH and swelling of the compartments. CQ (first named resochin) was developed in 1934 by Bayer laboratories as a synthetic antimalarial drug [113]. CQ was FDA-approved in 1949 and has proven to be one of the most effective and best tolerated agents against malaria [113,114]. It was widely used throughout the world in the 1950s and 1960s, and also later, but due to the emergence of the CQ-resistant malaria parasites *Plasmodium falciparum* and *Plasmodium vivax*, CQ has been abandoned as a prophylactic drug in most countries [115]. CQ and its analogues are also FDA-approved for the treatment of systemic lupus erythematosus and rheumatoid arthritis, and are in clinical trials as an adjuvant for anti-cancer chemotherapy and radiotherapy [115,116].

In short-term 3 h toxicity experiments in HEp-2 cells, a 2 h preincubation with 100 μM CQ gave a 15-fold increase in the Shiga toxin concentration needed to inhibit protein synthesis by 50% (IC50) [98]. In the retrograde pathway, the acidification decreases in the direction from early endosomes to the Golgi and the ER, and one might therefore expect CQ to have the most prominent effect early in the retrograde pathway. However, Shiga toxin transport to the Golgi apparatus was not affected by CQ treatment, neither was release of the A_1-moiety of the toxin in the ER, suggesting that transport to the ER was also normal [98]. Thus, CQ might interfere with StxA$_1$ translocation into the cytosol. Other compounds that disrupt the pH gradient, such as the V-ATPase inhibitors bafilomycin A1 (Baf) and concanamycin A (ConA) and the ionophore nigericin (Nig), also protect against Shiga toxin [98]. In contrast to CQ, the V-ATPase inhibitors interfere with Shiga toxin transport to the Golgi [98]. Even though all the compounds inhibit acidification, they interfere with different toxin transport steps, suggesting that also pH-independent processes are affected. The V-ATPase inhibitors show poor selectivity in vivo [117,118], and are therefore not suitable for clinical application.

Shiga toxin is shown to interact with the Sec61 channel [119,120], and the knockdown of Sec61B has been shown to protect cells against Shiga toxin [119], suggesting that the translocation of StxA$_1$ into the cytosol is dependent on the Sec61 channel. Interestingly, CQ has previously been shown to block several channels, such as the inward rectifier potassium channel Kir2.1 [121], and the translocation pores of the pore-forming toxins anthrax and C2 toxin [122,123]. One can speculate that CQ might also interact with the Sec61 channel to block the translocation of Shiga toxin. Importantly, as shown here, CQ also protects against Stx2 (see below).

For acute treatment of malaria, the recommended dose of CQ is 10 mg/kg/day. CQ is rapidly distributed in the body and reaches a peak plasma concentration within 3–12 h after an oral dose [124]. In most cases, CQ is slowly eliminated with a half-life median value of 40 days, but there are big differences between individuals and the half-life of CQ has been reported to range from 1 to 157 days [124]. The plasma concentration is strongly dependent on the administered dose and the duration of the treatment [124]. The therapeutic concentration in blood of hydroxychloroquine (HCQ), a less toxic analogue of CQ with similar pharmacokinetics, has been reported to be between 0.03 and 15 mg/L; the toxic and lethal concentration ranges were 3–26 mg/L and 20–104 mg/L, respectively [125]. CQ at high concentrations has been reported to have a number of effects in cell cultures, but clinically safe and achievable doses are in the low micromolar range [116]. Thus, it has been suggested that for clinical relevance, the CQ concentrations should not exceed 10 mg/L or 31 μM [116]. As shown in Figure 4, in long-term toxicity experiments (24 h with toxin), a 1 h preincubation with a concentration of 25 μM CQ gave approx. 20 fold protection against Stx2, indicating that CQ doses within the therapeutic window might be sufficient to protect against the toxin. HCQ also protected cells against Stx2 to a similar degree as CQ (Figure 4).

It should be noted that in animal models, CQ was shown to strongly accumulate in tissues, for instance in the uvea, liver, lungs, and kidneys [124]. A similar distribution has been reported in humans [124]. Thus, the CQ concentration in the target cells of Stx might be a lot higher than the plasma concentration.

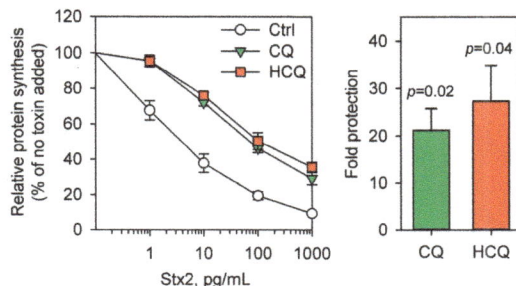

Figure 4. Cell protection against Stx2 by CQ and HCQ. HEp-2 cells were treated with or without 25 μM chloroquine (CQ) or 25 μM hydroxychloroquine (HCQ) in complete growth medium for 1 h prior to incubation with 10-fold serial dilutions of Stx2 for 24 h in the presence or absence of the drugs. The cells were then incubated in the presence of [³H]leucine for 20 min, and protein synthesis was measured as described in [98]. The left panel shows relative protein synthesis as a percentage of the samples without Stx2 added. The right panel shows relative fold protection against Stx2. The protection was calculated as an increase in IC50 for treated samples compared to control. The error bars show SEM (n = 4). One sample t-test was used for statistical analysis of the protection data, and obtained p values are given in the figure.

2.2. The Glucose Analogues 2-Deoxy-D-glucose and 2-Fluoro-2-deoxy-D-glucose

2-Deoxy-D-glucose (2DG) is a structural analogue of glucose, and has been used in research as a glycolytic inhibitor since 1950s [126,127]. 2DG differs from glucose only by the absence of one oxygen atom at the second carbon. Similarly to glucose, 2DG is taken up through the glucose transporters and phosphorylated by hexokinase to form 2DG-6-phosphate (2DG-6-P). However, 2DG-6-P is not metabolized further and accumulates in the cells [127–129]. 2DG-6-P inhibits glycolysis by competing with glucose-6-P for phosphoglucose isomerase [127], and by acting as a non-competitive inhibitor of hexokinase [130]. However, although the inhibition of glycolysis has been a commonly exploited effect of 2DG, this compound has a much broader spectrum of activities. In addition to inhibiting glycolysis, 2DG inhibits *N*-linked protein glycosylation [131,132]. In turn, this leads to accumulation of misfolded proteins and triggers the unfolded protein response in the ER, leading to ER stress [133,134]. Interestingly, Okuda et al. recently discovered that 2DG inhibits the expression of the Gb3 synthase by yet unknown mechanisms, and thus reduces cellular Gb3 levels in the cells [135].

2-Fluoro-2-deoxy-D-glucose (FDG) is a structural analogue of glucose where the hydroxyl group at the second carbon is replaced by a fluorine atom. Like glucose and 2DG, FDG is transported into cells, where it is phosphorylated by hexokinase to yield FDG-6-P. However, FDG-6-P does not undergo isomerization to fructose and thus cannot be further catabolized, leading to accumulation of FDG-6-P in the cells [136]. Similarly to 2DG, FDG also inhibits glycolysis, but because the binding energy of FDG-6-P for the allosteric site of the hexokinase is lower than that of 2DG-6-P, and closely resembles the energy of glucose-6-P, FDG is a better inhibitor of glycolysis than 2DG [137]. FDG also interferes with *N*-linked protein glycosylation [131,138,139], but in contrast to 2DG, FDG does not become incorporated into dolichol-linked oligosaccharides [138]. FDG seems to slow down rather than prevent the assembly of the dolichol-linked oligosaccharide, and thus is a weaker inhibitor of *N*-glycosylation than 2DG. In addition, we have recently found that in contrast to 2DG, FDG does not become incorporated into newly synthesized glycolipids [48]. [^{18}F]FDG, which contains the ^{18}F radioisotope, is a commonly used imaging agent for positron emission tomography (PET). [^{18}F]FDG based PET is widely used for diagnosis and monitoring of oncological, neurological, and cardiological diseases (for review see [140,141]).

We have recently discovered that both 2DG and FDG reduce cell sensitivity to both Shiga toxin and Stx2. Four hours preincubation with either 10 mM 2DG or 1 mM FDG led to an increase in the IC50 for Shiga toxin by 13-fold in HEp-2 cells [48,99]. In addition, 24 h preincubation with 10 mM 2DG reduced cell sensitivity to Shiga toxin by 30-fold [99], while 24 h pretreatment with 1 mM FDG made HEp-2 cells fully resistant to both Shiga toxin and Stx2 (maximal concentration tested was 100 ng/mL with 3 h challenge) [48]. For FDG, we have also tested whether the protective effect is observed in non-cancer originated cells, and found that immortalized human brain microvascular cells (HBMEC) also became less sensitive to Shiga toxin after 4 h and 24 h pretreatment with 1 mM FDG [48]. However, all these toxicity assays have been performed with a relatively short 3 h challenge with the toxin. Thus, we have now also tested whether FDG protects HEp-2 cells against 24 h challenge with Stx2, and observed an essentially similar protection as shown for 3 h incubation with the toxin (Figure 5).

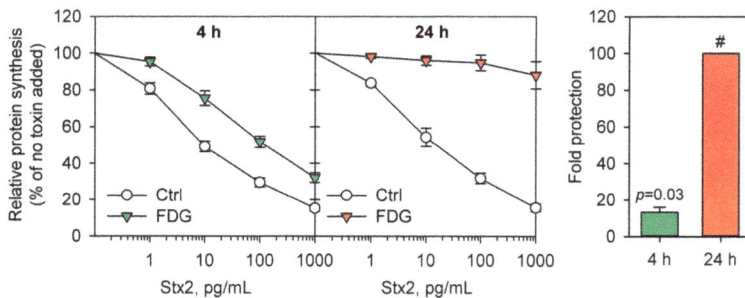

Figure 5. Cell protection against Stx2 by FDG. HEp-2 cells were treated with or without 1 mM 2-fluoro-2-deoxy-D-glucose (FDG) in complete growth medium for 4 h or 24 h prior to incubation with 10-fold serial dilutions of Stx2 for 24 h in the presence or absence of FDG. The cells were then incubated in the presence of [^{3}H]leucine for 20 min, and protein synthesis was measured as described in [98]. The left panel shows relative protein synthesis as a percentage of the samples without Stx2 added. The right panel shows relative fold protection against Stx2. The protection was calculated as an increase in IC50 for treated samples compared to control. In the samples treated with FDG for 24 h, the highest toxin concentration tested (1 ng/mL) did not reduce protein synthesis down to 50%, therefore the fold-protection could not be calculated and was estimated to be more than 100-fold (marked as #). The error bars show SEM for 4 h treatment (*n* = 4) and the deviation from the mean of two independent experiments for 24 h treatment. One sample *t*-test was used for statistical analysis of the protection data for 4 h treatment, and the obtained *p* value is given in the figure.

Interestingly, although both 2DG and FDG were found to reduce cellular Gb3 levels by 50% following 24 h treatment (Figure 6), it was only FDG that also led to reduction in Shiga toxin binding [48,99]. The mechanisms by which 2DG and FDG inhibit Gb3 synthesis are still not clear and seem to be different, as 2DG has been shown to inhibit the transcription of the Gb3 synthase gene [99,135], while FDG has no effect on the expression of Gb3 synthase [48]. In addition, we have found that cell treatment with FDG also reduces cellular levels of LacCer and glucosylceramide (GlcCer) (Figure 6) [48], indicating that FDG inhibits the synthesis of GlcCer and thus depletes cells for the precursors required for Gb3 synthesis, rather than inhibiting the synthesis of Gb3 directly.

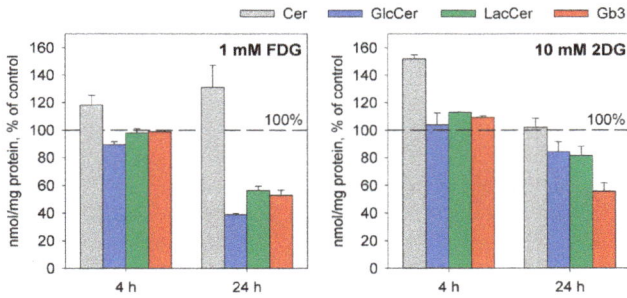

Figure 6. Effect of FDG and 2DG on total levels of Gb3 and its precursors. Cells were treated with or without 1 mM FDG or 10 mM 2DG for 4 h or 24 h, and lipids were analyzed by mass spectrometry in whole-cell lysates. The total amount of lipid was normalized to the total amount of protein in each sample (protein content was measured by BCA assay). The graph shows the levels of Cer, GlcCer, LacCer, and Gb3 in treated cells compared to control samples; the error bars show the deviation from the mean of two biological samples. For a detailed method description and the raw data see [48,99].

Although the inhibition of Gb3 synthesis might be an important factor for the protection against Stx by 2DG and FDG after long-term treatment, the protection observed after 4 h treatment with drugs does not seem to be mediated by changes in Gb3 [48,99]. We have shown that following 4 h treatment with either 10 mM 2DG or 1 mM FDG the intracellular transport of Shiga toxin is changed and most likely accounts for the protection observed at this time point. Four hours pretreatment with 10 mM 2DG almost completely blocked the release of Shiga toxin A_1-moeity in the ER in HEp-2 cells, which correlated well with the depletion of calcium from the ER [99]. Although it has been proposed that 2DG induces release of calcium from the ER via induction of ER stress [142], combined treatment with mannose, which rescues 2DG-mediated ER stress [143], does not prevent calcium leakage from the ER upon 2DG treatment and does not rescue cell sensitivity to Shiga toxin [99]. Three ER chaperones, HEDJ (also called ERdj3), BiP (also called GRP78) and GRP94 (glucose-regulated protein of 94 kDa), have been shown to bind the A-moiety of Stx1 [120], and thus are suggested to be involved in the release of StxA₁ in the ER. Substrate binding to GRP94 and BiP has been shown to be regulated by calcium [144,145], suggesting that 2DG might prevent the release of StxA₁ in the ER by inhibiting Shiga toxin interaction with chaperones.

Similar to 2DG, FDG was found to deplete calcium from the ER and to inhibit release of the A_1-moiety from the holotoxin [48]. However, since the concentrations for FDG and 2DG used were different, it is difficult to conclude whether the effect on ER calcium levels and the efficiency in blocking StxA₁ release are similar for these drugs. However, we have found that FDG also inhibits Stx1 transport from the Golgi to ER [48], which did not seem to be the case for 2DG [99]. This additional block in the Stx transport might explain why FDG is more efficient than 2DG in protecting cells against Shiga toxin and Stx2 following short-term incubation. Furthermore, since the long-term preincubation (24 h) with FDG, but not 2DG, also leads to reduced Stx1 binding, it makes FDG a promising candidate drug for STEC infections. However, in vivo studies are required to test whether FDG could potentially be used for treatment. The first challenge is to reach high enough concentrations of FDG in the Stx-targeted cells. When [18F]FDG is used in the clinics for PET, the main limiting factor for the concentration of [18F]FDG used in patients is the allowed maximal radiation dose. This is not the problem when using stable nonradioactive FDG. There are no clinical studies describing maximal FDG doses that could be safely achieved in human plasma/tissues, but there are several clinical studies that have analyzed the safety of 2DG when used in combination with chemotherapy or radiation. Based on the study by Raez et al. [146] the recommended daily dose of 2DG in combination with docetaxel was 63 mg/kg, which resulted in a median maximum plasma 2DG concentration of 0.7 mM, and caused tolerable adverse effects. The tolerable concentration and adverse effects of FDG might differ from those shown for 2DG and have to be assessed in the future, but if the achievable concentration is similar to that shown for 2DG (0.7 mM), it is then in the same range as the concentration (1 mM) shown to protect cells against Stx in vitro [48].

2.3. Retro-2 Substances

By performing a high-throughput screening of 16,480 drugs Stechmann et al. [100] identified two low molecular weight substances that reduced sensitivity to Stx1, Stx2, and the plant toxin ricin (which also follows a retrograde route to the ER), when added 30 min prior to challenge with toxins in A459 and HeLa cells. These substances were called Retro-1 and Retro-2. They were reported to inhibit retrograde transport of Stx1B from endosomes to the Golgi apparatus without affecting compartment integrity and endogenous retrograde cargo transport [100]. Other compounds, which may have similarities to the Retro compounds, have also been shown to block trafficking and toxicity of Stx1 and ricin [147]. Both Retro-1 and Retro-2 have been found to relocalize the SNARE proteins syntaxin 5 and, to a lesser extent, syntaxin 6 from their normal site of accumulation on perinuclear Golgi membranes [100]. Thus, it has been suggested that the inhibition of Stx1B transport by these compounds could be mediated by the relocalization of syntaxin 5 and 6, although additional studies are required to confirm this. Retro-2 was found to be the most effective of these drugs, and an intraperitoneal injection of 200 mg/kg Retro-2 given 1 h prior to toxin challenge completely protected mice that were given a lethal nasal instillation of ricin (animal experiments with Stx were not performed) [100]. Later, this group and others published data for several substances similar to Retro-2, and showed that it was a cyclic form (Retro-2^{cycl}) and not Retro-2 that was active [148–150]. Cyclization and modification of Retro-2 resulted in a compound with approx. 100-fold increased efficacy in inhibiting Stx1, and only one enantiomer was found to be active with an EC50 value (the concentration of the drug that gives 50% of its full inhibitory effect against the toxin) of approx. 300 nM in HeLa cells [149]. The most active of the Retro substances in counteracting the cytotoxicity of Stx1 (named (S)-Retro-2.1) has been reported to have an EC50 value of 54 nM in HeLa cells [151].

Secher et al. has investigated whether Retro-2^{cycl} could protect mice against the toxic effects of infection with *E. coli* O104:H4, the strain that was responsible for the deadly outbreak in Europe in 2011 [152,153]. The bacteria were given to mice by oral gavage, and on days 16 and 26 the mice received intraperitoneal injections of 100 mg/kg of Retro-2^{cycl}, which resulted in reduced mortality rate in the treated group (29 and 16 mice were dead in control and treated group, respectively, out of 40 mice per group) [101]. The same authors recently published a review article about the use of Retro-2 and similar compounds to protect against Stxs, ricin, multiple viruses, including different polyomaviruses, Ebola virus and poxviruses, and intracellular parasites, such as *Leishmania* [154], and we refer to this review for further discussions about these substances. The authors conclude that these lead compounds now need to be developed as drugs for human use. As these drugs, as far as we know, have not been given to humans or even been tested in formal preclinical drug safety studies, that task will be ongoing for many years. Furthermore, it will be interesting to see which compound will be selected for such a development as the most efficient compound may also be the most toxic. Finally, it should also be investigated whether all of these compounds have similar effects on the localization of syntaxin 5 as Retro-1 and Retro-2 [100], since targeting of a universal trafficking factor such as syntaxin 5 may prove challenging in a clinical setting.

2.4. Manganese

Mn^{2+}-ions have been described by several groups to protect cells against Shiga toxin [105] and Stx1 [103]. Mukhopadhyay and Lindstedt reported that the protection was due to redirection of the toxin to the degradative pathway; they also reported that mice injected with a lethal dose of Stx1 could be rescued by a nontoxic dose of Mn^{2+} that was injected five days before the toxin [103]. However, Mn^{2+} failed to protect cells against Stx2 under the same conditions [65]. These authors also showed that elevated levels of Mn^{2+} resulted in a down-regulation of GPP130, which in turn led to inhibition of retrograde transport of Stx1, but not Stx2 [65]. Furthermore, this group later showed that the increased Mn^{2+} levels induced GPP130 oligomerization and its sorting to lysosomes for degradation [102].

The protective effect of Mn^{2+} against Stx toxicity was also studied by another group, and there the authors failed to see any protective effect of Mn^{2+} against Stx1 and Stx2 in cultured Vero cells and CD-1 outbred mice at Mn^{2+}-doses that were not toxic to the cells and the animals [104]. They concluded that the ability of Mn^{2+} to protect against Stx toxicity might be dependent on the cell line and mouse strain, and that protection may be observed only at potentially toxic concentrations of Mn^{2+}. These authors and others have discussed that Mn^{2+}-ions are neurotoxic; for reviews see [155,156]. Due to this well-known toxic effect of Mn^{2+}, we believe that it is unlikely that Mn^{2+} can be developed as a drug against Stx toxicity.

2.5. Inhibitors of GlcCer Synthesis PDMP and C-9

The glycosphingolipid Gb3 is the sole functional receptor for Stx in humans, which makes it a potential target for preventing Stx toxicity to cell. The drawback of the approach directed towards the receptor is the time required to deplete Gb3 from cells, and this might limit the therapeutic potential. However, although the expression of Gb3 is a prerequisite for cell sensitivity to Stx, a specific Gb3 species composition [42–45] is required for efficient Stx binding and intracellular transport, meaning that a complete depletion of Gb3 might not be necessary to prevent Stx intoxication. To our knowledge, there are no compounds available that would specifically block Gb3 synthesis in the cells. However, multiple substances have been developed for inhibiting glucosylceramide synthesis, and thus prevent formation of more complex glycosphingolipids, including Gb3. Two such compounds, PDMP [43] and C-9 [106], have been shown to reduce cell sensitivity to Stx, and are discussed in this section.

PDMP (1-phenyl-2-decanoyl-amino-3-morpholino-1-propanol) is a ceramide analogue first developed in a search for drugs to treat individuals with Gaucher disease [157], which are deficient of the lysosomal enzyme glucosylceramidase and thus accumulate GlcCer in certain tissues [158]. PDMP has been shown to inhibit the synthesis of GlcCer and specifically affect the content and composition of glycosphingolipids, without perturbing other lipid profiles [43,159], and without significantly affecting the synthesis of glycoproteins in cells [159]. Raa et al. showed that treatment with 1 µM PDMP for 24 h had only a small (20%–30%) effect on Shiga toxin binding to HEp-2 cells, but led to an approx. 50% reduction in the toxin uptake and almost completely blocked (by 90%) the transport of StxB into the Golgi [43]. These effects were accompanied by a 6.5-fold increase in the IC50 value for Shiga toxin in HEp-2 cells. After 24 h incubation with 1 µM PDMP, the cellular levels of Gb3 and its precursors were reduced by approx. 50% in treated HEp-2 cells, and the Gb3 species with the fatty acyl group 16:0 were found to be degraded faster and to a larger extent than the Gb3 with the fatty acid 24:1 [43]. Results from studies on butyric acid-mediated cell sensitization to Shiga toxin, and comparison of cell lines with different sensitivities to Stx, have indicated that Gb3 with certain fatty acyl groups might be important for endosome-to-Golgi transport [49,91,160,161]. Thus, the changes in Gb3 species composition may play an important role in the PDMP-induced protection against Stx, at least at shorter treatment times, when the total Gb3 is not yet completely depleted from the cells. However, in addition to GlcCer synthase, PDMP has been found to target other lipid enzymes, such as ceramide glycanase and a lysosomal phospholipase A2 called 1-O-acylceramide synthase [162,163], which might limit the applicability of PDMP for specific depletion of Gb3 in vivo. On the other hand, the development of PDMP has boosted the synthesis of a variety of related compounds [162]. Some of these PDMP analogues exhibit a dramatic increase in potency and specificity for the ceramide-specific glucosyl transferase [162], and have been tested in β-galactosidase a-null mice (model of Fabry disease in which Gb3 accumulates in the vasculature and kidneys) [164]. The mice were injected intraperitoneally with 2 mg/kg of the PDMP analogue EtDO-P4 (D-threo-1-(3,4-ethylenedioxyphenyl)-2-(palmitoylamino)-3-(1-pyrrolidinyl)propanol) twice a day for three days, which led to approx. 50% reduction in GlcCer levels in the kidney, liver, and heart.

The reduction in Gb3 was little pronounced after three days of treatment, but following eight weeks of treatment with 10 mg/kg of EtDO-P4 twice a day, the total levels of Gb3 were reduced by approx. 50% in the kidney, liver and heart. Importantly, the treatment did not show apparent toxicity to the animals. However, a potential protection against Stx toxicity was not tested in this study.

The search for treatment of glycosphingolipid storage diseases has led to the development of another ceramide analogue named C-9 [(1R, 2R)-nonanoic acid [2-(2′,3′-dihydro-benzo [1–4]dioxin-6′-yl)-2-hydroxy-1-pyrrolidin-1-ylmethyl-ethyl]-amide-L-tartaric acid salt] (Genzyme Corp., Waltham, MA, USA). Silberstein et al. has demonstrated that a 48 h pretreatment with 5 µM C-9 reduced cellular Gb3 levels by approx. 80% in human renal tubular epithelial cells (HRTEC), which was accompanied by an almost complete cell protection against a 24 h challenge with 1 ng/mL Stx2 [106]. However, the drug had no effect on the cell sensitivity to Stx2 when added at the same time as the toxin and a 24 h preincubation was required to obtain the protective effect [106]. Essentially similar protection and reduction in Gb3 was also shown in human glomerular endothelial cells [108]. Silberstein et al. also investigated the potency of C-9 against Stx2 in an in vivo model [107]. Rats received C-9 orally two days prior and four days after the intraperitoneal injection of the supernatant from recombinant *E. coli* expressing Stx2. The treatment reduced rat mortality by 50% and prevented intestinal and renal tissue damage, which was observed in the group treated with Stx2 only. The failure of C-9 to completely prevent rat mortality after Stx2 challenge was attributed to the possibility that C-9 does not pass the blood- brain barrier, and thus the deaths in the C-9 group could be the outcome of neurological injuries. However, no histological analysis was performed to support this [107].

A related compound, which differs from C-9 only in the fatty acid part (contains octanoic acid instead of nonanoic acid), has been approved by FDA for treatment of Gaucher disease [165] and is now sold under the name Cerdelga® (Eliglustat tartrate; Genzyme Corp., Cambridge, MA, USA). Eliglustat tartrate (also called Genz-112638) is well tolerated and has a recommended dosing of 100 mg (contains 84 mg of eliglustat) twice daily in Gaucher patients [166,167]. When used at the recommended dosing, the average concentration of free eliglustat in the plasma is 12–25 ng/mL (but there is a great variation between individuals due to differences in the rate of metabolic degradation of eliglustat) (reviewed in [165]). Thus, it would be interesting to see whether eliglustat provides similar protection against Stx in cells and in animal models as C-9, and whether the tolerated dose leads to significant changes in Gb3 levels and/or species, which would indicate its potential use for the treatment of STEC infections. However, very little eliglustat is taken up into brain [164], which would limit its effect in the HUS cases with neuronal damage.

2.6. HG

The alkylglycerol 1-O-hexadecyl-*sn*-glycerol (HG, also called chimyl alcohol) is a precursor for ether-linked glycerophospholipids. HG enters the biosynthetic pathway of ether lipids after being phosphorylated by alkylglycerol kinase [168]. Addition of HG was shown to alter the lipidome of HEp-2 cells, with the most notable changes being an increase in the cellular level of ether-linked lipids with 16 carbon atoms at *sn*-1 position, an increase in lysophosphatidylinositol (LPI) and a decrease in all glycosphingolipid classes analyzed, amongst them the Stx receptor Gb3 [50].

We recently showed that a 24 h preincubation with 20 µM HG strongly protected HEp-2, HMEC-1 and HBMEC cells against Shiga toxin and Stx2, with an average 30-fold increase of IC50 [50]. There was a moderate reduction in Stx1 binding to HEp-2 cells after HG treatment, presumably due to reduced Gb3 levels, but this decrease was not sufficiently large to account for the strong protection against the toxin. Immunofluorescence confocal microscopy revealed that HG treatment led to an accumulation of Stx1 in the Golgi apparatus after 4 h of toxin challenge, while in control cells, the toxin clearly colocalized with both the ER-marker PDI and the Golgi-marker TGN46 at the same time point, suggesting that HG treatment inhibits Golgi-to-ER transport of Stx [50]. Thus, HG prevents Stx intoxication primarily by interfering with the Golgi-to-ER transport of Stx, but also to some degree by downregulating the Stx receptor Gb3.

The 1-*O*-alkylglycerols, including HG, are naturally occurring ether lipids that are present in human cells and body fluids, particularly in hematopoietic organs and in neutrophils and human milk [169]. Alkylglycerols can also be obtained from the diet, with marine oils, especially shark liver oil, being a good source [169,170]. Shark liver oil contains 9%–13% HG and has been used in traditional medicine in the Nordic countries for beneficial health effects and for wound healing [169,170]. Shark liver oil and alkylglycerol supplementation have been reported to mediate several biological effects, including the ability to boost the immune system and alleviate radiation therapy-induced side effects [169,170]. However, to evaluate the potential of HG as an inhibitor of Stx intoxication, further studies, first in animal models, are required.

2.7. Statins

Statins, also known as HMG-CoA (3-hydroxy-3-methylglutaryl coenzyme A) reductase inhibitors, are a class of compounds that inhibit cholesterol biosynthesis and prenylation of proteins (reviewed in [171]). Statins are widely prescribed to lower serum cholesterol levels for the prevention of cardiac diseases [172–174]. Recently, statins were found to inhibit Stx1B transport to the Golgi apparatus and to protect cells against Stx1 and Stx2 [110]. ACHN (epithelial carcinoma from renal tubular adenocarcinoma pleural metastasis) cells were pretreated with rosuvastatin (5–20 μM) for 24 h prior to addition of Stx1 or Stx2, and the number of surviving cells was measured after 72 h. Treatment with rosuvastatin increased cell survival by approx. 25% in the group treated with 5 μM rosuvastatin and then challenged with 1 pg/mL of Stx1 or Stx2, but there was essentially no improvement in cell survival in the group challenged with 100 pg/mL of the toxins. On the other hand, cell treatment with 10 μM of rosuvastatin increased cell survival by approx 70% in the groups challenged with 1 pg/mL of Stx1 or Stx2, and by approx. 25% in the groups challenged with100 pg/mL of the toxins. Although increasing the concentrations of rosuvastatin to 20 μM showed even better protection, there seemed to appear a difference in protection against Stx1 and Stx2, with the protection against Stx2 being lower than that against Stx1, especially at high toxin doses. As previously mentioned, Stx2 is suggested to exploit an additional transport pathway [65] which might not be blocked by rosuvastatin. In addition, a protective effect by lovastatin has previously been shown for several other protein toxins, including ricin, modeccin, *Pseudomonas* toxin, and diphtheria toxin, suggesting that statins might have a potential to be used against several toxins [175].

Cell treatment with statins was found to increase the protein level and activity of GlcCer synthase, which in turn led to increased cellular levels of GlcCer [110]. The fact that the cells were maintained in a medium containing cholesterol during the treatment with statins, and that the isoprenol precursor counteracted the upregulation of GlcCer synthesis induced by statins, indicated that the effects observed, at least for the upregulation of GlcCer synthesis, were mediated by the inhibition of isoprenylation rather than by depletion of cellular cholesterol. In addition, by the use of specific inhibitors of different prenylation enzymes, the authors were able to show that it was only the inhibition of geranylgeranyl transferase II, also known as Rab geranylgeranyl transferase, that led to similar upregulation of GlcCer as statin treatment. Rab prenylation facilitates their membrane association and activity [176,177]. Rab GTPases regulate intracellular vesicular trafficking events, and several Rabs have also been implicated in Stx retrograde transport (reviewed in [1]), suggesting that rosuvastatin-induced protection against Stx is mediated via aberrant Rab prenylation.

Importantly, although for the prevention of cholesterol-related cardiac diseases statins are prescribed in relatively low doses (recommended daily dose of lovastatin is 20–40 mg per day [174]) the achievable safe serum values might be as high as 0.2–12 μM (shown for lovastatin at doses of 133–412 mg/m^2, which corresponds to approx. 200–650 mg) [178], and are within the dose range found to protect cells against Stx1 and Stx2 [110].

2.8. Furin Inhibitors

Furin is a type I transmembrane serine protease that activates precursors of different physiologically important proteins. Although furin is mainly located in the Golgi and *trans*-Golgi network, it also circulates through the endosomal system to the cell surface and back to the Golgi [179], and may also be secreted as a soluble truncated active enzyme [180]. In addition to its physiological role, furin activates numerous toxic proteins, including Stxs [95]. Moreover, furin has been implicated in various human diseases, including cancer, osteoarthritis, atherosclerosis, diabetes and neurodegenerative disorders [181–184]. Although the complete knockout of the *fur* gene, which encodes furin, is embryonically lethal to mice, the specific inhibition of furin by polyarginine inhibitors has been shown to be well tolerated in adult animals [185,186], indicating that short term inhibition of furin might be exploited for treatment. Thus, various furin inhibitors have been developed and investigated for their potential therapeutic applications during the last years [111,112,183,184,187]. For instance, the furin inhibitor hexa-D-arginine amide has been demonstrated to improve the survival of mice challenged with *Pseudomonas aeruginosa* exotoxin A, which requires furin-mediated cleavage for toxicity [185]. Later, the same compound was shown to delay anthrax toxin-induced toxemia both in cells and in rats [186], supporting the therapeutic potential of furin inhibitors for the treatment of infectious diseases.

A number of peptidomimetic furin inhibitors have been developed and optimized to improve the activity and stability of the compounds [111], and some of these compounds have been found to protect against Shiga toxin without showing significant toxicity to cells [111]. The most effective analogue for the protection against Shiga toxin was 4-(guanidinomethyl)phenylacetyl-Arg-Val-Arg-4-amidinobenzylamide (called No. 24 in the original article), which reduced HEp-2 cells sensitivity to Shiga toxin by approx. 6-fold, when added to the cells 30 min before a 4 h incubation with unnicked Shiga toxin [111]. However, it has not yet been investigated whether cells are protected against longer challenges with the toxin. It has been shown that the Shiga toxin may be processed by other cellular proteases, but more slowly [39,95]. Thus, it is still not clear whether furin inhibitors help to prevent or treat HUS in humans infected with Stx-producing bacteria.

3. Concluding Remarks

A number of compounds that protect cells against Stxs are known, and in some cases they have also been found to protect animals against the challenge with purified toxin or toxic effects of infection with STEC. Most of these compounds have not been used in humans and thus need to be carefully evaluated for potential use in the clinic, and it is not clear to which extent one will be able to find compounds that are safe in humans. There are, however, some exceptions, for instance chloroquine and statins, which should be investigated for a possible protective effect in connection with Stx-induced disease. In addition, there is an ongoing search for efficient inhibitors of glycosphingolipid synthesis for the treatment of glycosphingolipid storage diseases [188] as well as cancer [189], meaning that new, more potentially useful compounds will be investigated in the near future. Such new compounds should not be overlooked for their potential in the treatment of STEC and *Shigella* infections.

Acknowledgments: The work has been supported by the Norwegian Cancer Society, the Research Council of Norway through its Centers of Excellence funding scheme, project number 179571, and the South-Eastern Norway Regional Health Authority.

Conflicts of Interest: The authors declare no conflict of interest.

Abbreviations

ACHN	epithelial carcinoma from renal tubular adenocarcinoma pleural metastasis
Baf	bafilomycin A1
ConA	concanamycin A
cPLA2	cytoplasmic phospholipase A2

CQ	chloroquine
C-9	(1R, 2R)-nonanoic acid [2-(2′,3′-dihydro-benzo [1–4]dioxin-6′-yl)-2-hydroxy-1-pyrrolidin-1-ylmethyl-ethyl]-amide-L-tartaric acid salt
EC50	the concentration of the drug that gives 50% of its full inhibitory effect against the toxin
FDG	2-fluoro-2-deoxy-D-glucose
Gal	galactose
Gb3	globotriaosylceramide
Glc	glucose
GlcCer	glucosylceramide
GPP130	Golgi phosphoprotein of 130 kDa
HBMEC	human brain microvascular endothelial cells
HCQ	hydroxychloroquine
HG	alkylglycerol 1-O-hexadecyl-*sn*-glycerol
HRTEC	human renal tubular epithelial cells
HUS	hemolytic-uremic syndrome
HuSAP	human serum amyloid component P
IC50	Stx concentration needed to inhibit protein synthesis by 50%
LacCer	lactosylceramide
LPI	lysophosphatidylinositol
Nig	nigericin
PDMP	1-phenyl-2-decanoyl-amino-3-morpholino-1- propanol
PET	positron emission tomography
Retro-2^{cycl}	cyclic form of Retro-2
STEC	Shiga toxin-producing *E. coli*
Stx	Shiga toxins (when referring to the whole family and common features)
StxB	B-moiety of Stx
Stx1	Shiga-like toxin 1
Stx2	Shiga-like toxin 2
TLC	thin layer chromatography
UDP	uridine diphosphate
[^{18}F]FDG	FDG which contains the ^{18}F radioisotope
2DG	2-deoxy-D-glucose

References

1. Bergan, J.; Dyve Lingelem, A.B.; Simm, R.; Skotland, T.; Sandvig, K. Shiga toxins. *Toxicon* **2012**, *60*, 1085–1107. [CrossRef] [PubMed]
2. Scheutz, F.; Teel, L.D.; Beutin, L.; Piérard, D.; Buvens, G.; Karch, H.; Mellmann, A.; Caprioli, A.; Tozzoli, R.; Morabito, S.; et al. Multicenter evaluation of a sequence-based protocol for subtyping Shiga toxins and standardizing Stx nomenclature. *J. Clin. Microbiol.* **2012**, *50*, 2951–2963. [CrossRef] [PubMed]
3. Tesh, V.L.; Burris, J.A.; Owens, J.W.; Gordon, V.M.; Wadolkowski, E.A.; O'Brien, A.D.; Samuel, J.E. Comparison of the relative toxicities of Shiga-like toxins type I and type II for mice. *Infect. Immun.* **1993**, *61*, 3392–3402. [PubMed]
4. Siegler, R.L.; Obrig, T.G.; Pysher, T.J.; Tesh, V.L.; Denkers, N.D.; Taylor, F.B. Response to Shiga toxin 1 and 2 in a baboon model of hemolytic uremic syndrome. *Pediatr. Nephrol.* **2003**, *18*, 92–96. [PubMed]
5. Karch, H.; Tarr, P.I.; Bielaszewska, M. Enterohaemorrhagic *Escherichia coli* in human medicine. *Int. J. Med. Microbiol.* **2005**, *295*, 405–418. [CrossRef] [PubMed]
6. Palermo, M.S.; Exeni, R.A.; Fernandez, G.C. Hemolytic uremic syndrome: pathogenesis and update of interventions. *Expert Rev. Anti Infect. Ther.* **2009**, *7*, 697–707. [CrossRef] [PubMed]
7. Agger, M.; Scheutz, F.; Villumsen, S.; Molbak, K.; Petersen, A.M. Antibiotic treatment of verocytotoxin-producing *Escherichia coli* (VTEC) infection: a systematic review and a proposal. *J. Antimicrob. Chemother.* **2015**, *70*, 2440–2446. [CrossRef] [PubMed]
8. Barton Behravesh, C.; Jones, T.F.; Vugia, D.J.; Long, C.; Marcus, R.; Smith, K.; Thomas, S.; Zansky, S.; Fullerton, K.E.; Henao, O.L.; et al. Deaths associated with bacterial pathogens transmitted commonly

through food: foodborne diseases active surveillance network (FoodNet), 1996–2005. *J. Infect. Dis.* **2011**, *204*, 263–267. [CrossRef] [PubMed]

9.   Boisen, N.; Melton-Celsa, A.R.; Scheutz, F.; O'Brien, A.D.; Nataro, J.P. Shiga toxin 2a and Enteroaggregative *Escherichia coli*—a deadly combination. *Gut Microbes* **2015**, *6*, 272–278. [CrossRef] [PubMed]

10. Tarr, P.I.; Gordon, C.A.; Chandler, W.L. Shiga-toxin-producing *Escherichia coli* and haemolytic uraemic syndrome. *Lancet* **2005**, *365*, 1073–1086. [CrossRef]

11. Gould, L.H.; Demma, L.; Jones, T.F.; Hurd, S.; Vugia, D.J.; Smith, K.; Shiferaw, B.; Segler, S.; Palmer, A.; Zansky, S.; et al. Hemolytic uremic syndrome and death in persons with *Escherichia coli* O157:H7 infection, foodborne diseases active surveillance network sites, 2000–2006. *Clin. Infect. Dis.* **2009**, *49*, 1480–1485. [CrossRef] [PubMed]

12. Brandt, J.R.; Fouser, L.S.; Watkins, S.L.; Zelikovic, I.; Tarr, P.I.; Nazar-Stewart, V.; Avner, E.D. *Escherichia coli* O157:H7-associated hemolytic-uremic syndrome after ingestion of contaminated hamburgers. *J. Pediatr.* **1994**, *125*, 519–526. [CrossRef]

13. Hughes, D.A.; Beattie, T.J.; Murphy, A.V. Haemolytic uraemic syndrome: 17 years' experience in a Scottish paediatric renal unit. *Scott. Med. J.* **1991**, *36*, 9–12. [CrossRef] [PubMed]

14. Taylor, C.M.; White, R.H.; Winterborn, M.H.; Rowe, B. Haemolytic-uraemic syndrome: Clinical experience of an outbreak in the West Midlands. *Br. Med. J. (Clin. Res. Ed.)* **1986**, *292*, 1513–1516. [CrossRef]

15. Bale, J.F., Jr.; Brasher, C.; Siegler, R.L. CNS manifestations of the hemolytic-uremic syndrome. Relationship to metabolic alterations and prognosis. *Am. J. Dis. Child.* **1980**, *134*, 869–872. [CrossRef] [PubMed]

16. Obata, F.; Tohyama, K.; Bonev, A.D.; Kolling, G.L.; Keepers, T.R.; Gross, L.K.; Nelson, M.T.; Sato, S.; Obrig, T.G. Shiga toxin 2 affects the central nervous system through receptor globotriaosylceramide localized to neurons. *J. Infect. Dis.* **2008**, *198*, 1398–1406. [CrossRef] [PubMed]

17. Armstrong, G.D.; Fodor, E.; Vanmaele, R. Investigation of Shiga-like toxin binding to chemically synthesized oligosaccharide sequences. *J. Infect. Dis.* **1991**, *164*, 1160–1167. [CrossRef] [PubMed]

18. Trachtman, H.; Cnaan, A.; Christen, E.; Gibbs, K.; Zhao, S.; Acheson, D.W.; Weiss, R.; Kaskel, F.J.; Spitzer, A.; Hirschman, G.H. Effect of an oral Shiga toxin-binding agent on diarrhea-associated hemolytic uremic syndrome in children: a randomized controlled trial. *JAMA* **2003**, *290*, 1337–1344. [CrossRef] [PubMed]

19. Nishikawa, K.; Matsuoka, K.; Kita, E.; Okabe, N.; Mizuguchi, M.; Hino, K.; Miyazawa, S.; Yamasaki, C.; Aoki, J.; Takashima, S.; et al. A therapeutic agent with oriented carbohydrates for treatment of infections by Shiga toxin-producing *Escherichia coli* O157:H7. *Proc. Natl. Acad. Sci. USA* **2002**, *99*, 7669–7674. [CrossRef] [PubMed]

20. Mulvey, G.L.; Marcato, P.; Kitov, P.I.; Sadowska, J.; Bundle, D.R.; Armstrong, G.D. Assessment in mice of the therapeutic potential of tailored, multivalent Shiga toxin carbohydrate ligands. *J. Infect. Dis.* **2003**, *187*, 640–649. [CrossRef] [PubMed]

21. Melton-Celsa, A.R.; Carvalho, H.M.; Thuning-Roberson, C.; O'Brien, A.D. Protective efficacy and pharmacokinetics of human/mouse chimeric anti-Stx1 and anti-Stx2 antibodies in mice. *Clin. Vaccine Immunol.* **2015**, *22*, 448–455. [CrossRef] [PubMed]

22. Kimura, T.; Tani, S.; Matsumoto Yi, Y.; Takeda, T. Serum amyloid P component is the Shiga toxin 2-neutralizing factor in human blood. *J. Biol. Chem.* **2001**, *276*, 41576–41579. [CrossRef] [PubMed]

23. Armstrong, G.D.; Mulvey, G.L.; Marcato, P.; Griener, T.P.; Kahan, M.C.; Tennent, G.A.; Sabin, C.A.; Chart, H.; Pepys, M.B. Human serum amyloid P component protects against *Escherichia coli* O157:H7 Shiga toxin 2 in vivo: Therapeutic implications for hemolytic-uremic syndrome. *J. Infect. Dis.* **2006**, *193*, 1120–1124. [CrossRef] [PubMed]

24. Kielstein, J.T.; Beutel, G.; Fleig, S.; Steinhoff, J.; Meyer, T.N.; Hafer, C.; Kuhlmann, U.; Bramstedt, J.; Panzer, U.; Vischedyk, M.; et al. Best supportive care and therapeutic plasma exchange with or without eculizumab in Shiga-toxin-producing *E. coli* O104:H4 induced haemolytic-uraemic syndrome: an analysis of the German STEC-HUS registry. *Nephrol. Dial. Transplant.* **2012**, *27*, 3807–3815. [CrossRef] [PubMed]

25. Orth, D.; Khan, A.B.; Naim, A.; Grif, K.; Brockmeyer, J.; Karch, H.; Joannidis, M.; Clark, S.J.; Day, A.J.; Fidanzi, S.; et al. Shiga toxin activates complement and binds factor H: evidence for an active role of complement in hemolytic uremic syndrome. *J. Immunol.* **2009**, *182*, 6394–6400. [CrossRef] [PubMed]

26. Melton-Celsa, A.R.; O'Brien, A.D. New therapeutic developments against Shiga toxin-producing *Escherichia coli*. *Microbiol. Spectr.* **2014**, *2*. [CrossRef] [PubMed]

27. Stein, P.E.; Boodhoo, A.; Tyrrell, G.J.; Brunton, J.L.; Read, R.J. Crystal structure of the cell-binding B oligomer of verotoxin-1 from *E. coli*. *Nature* **1992**, *355*, 748–750. [CrossRef] [PubMed]

28. Fraser, M.E.; Chernaia, M.M.; Kozlov, Y.V.; James, M.N. Crystal structure of the holotoxin from *Shigella dysenteriae* at 2.5 Å resolution. *Nat. Struct. Biol.* **1994**, *1*, 59–64. [CrossRef] [PubMed]

29. Jacewicz, M.; Clausen, H.; Nudelman, E.; Donohue-Rolfe, A.; Keusch, G.T. Pathogenesis of shigella diarrhea. XI. Isolation of a shigella toxin-binding glycolipid from rabbit jejunum and HeLa cells and its identification as globotriaosylceramide. *J. Exp. Med.* **1986**, *163*, 1391–1404. [CrossRef] [PubMed]

30. Lindberg, A.A.; Brown, J.E.; Stromberg, N.; Westling-Ryd, M.; Schultz, J.E.; Karlsson, K.A. Identification of the carbohydrate receptor for Shiga toxin produced by *Shigella dysenteriae* type 1. *J. Biol. Chem.* **1987**, *262*, 1779–1785. [PubMed]

31. Lingwood, C.A.; Law, H.; Richardson, S.; Petric, M.; Brunton, J.L.; De, G.S.; Karmali, M. Glycolipid binding of purified and recombinant *Escherichia coli* produced verotoxin *in vitro*. *J. Biol. Chem.* **1987**, *262*, 8834–8839. [PubMed]

32. DeGrandis, S.; Law, H.; Brunton, J.; Gyles, C.; Lingwood, C.A. Globotetraosylceramide is recognized by the pig edema disease toxin. *J. Biol. Chem.* **1989**, *264*, 12520–12525. [PubMed]

33. Ling, H.; Boodhoo, A.; Hazes, B.; Cummings, M.D.; Armstrong, G.D.; Brunton, J.L.; Read, R.J. Structure of the Shiga-like toxin I B-pentamer complexed with an analogue of its receptor Gb3. *Biochemistry* **1998**, *37*, 1777–1788. [CrossRef] [PubMed]

34. Peter, M.G.; Lingwood, C.A. Apparent cooperativity in multivalent verotoxin-globotriaosyl ceramide binding: kinetic and saturation binding studies with [^{125}I]verotoxin. *Biochim. Biophys Acta* **2000**, *1501*, 116–124. [CrossRef]

35. Bast, D.J.; Banerjee, L.; Clark, C.; Read, R.J.; Brunton, J.L. The identification of three biologically relevant globotriaosyl ceramide receptor binding sites on the Verotoxin 1 B subunit. *Mol. Microbiol.* **1999**, *32*, 953–960. [CrossRef] [PubMed]

36. Marcato, P.; Mulvey, G.; Armstrong, G.D. Cloned Shiga toxin 2 B subunit induces apoptosis in Ramos Burkitt's lymphoma B cells. *Infect. Immun.* **2002**, *70*, 1279–1286. [CrossRef] [PubMed]

37. Kymre, L.; Simm, R.; Skotland, T.; Sandvig, K. Different roles of the C-terminal end of Stx1A and Stx2A for AB$_5$ complex integrity and retrograde transport of Stx in HeLa cells. *Pathog. Dis.* **2015**, *73*, ftv083. [CrossRef] [PubMed]

38. Fraser, M.E.; Fujinaga, M.; Cherney, M.M.; Melton-Celsa, A.R.; Twiddy, E.M.; O'Brien, A.D.; James, M.N. Structure of Shiga toxin type 2 (Stx2) from *Escherichia coli* O157:H7. *J. Biol. Chem.* **2004**, *279*, 27511–27517. [CrossRef] [PubMed]

39. Garred, O.; van Deurs, B.; Sandvig, K. Furin-induced cleavage and activation of Shiga toxin. *J. Biol. Chem.* **1995**, *270*, 10817–10821. [PubMed]

40. Tam, P.J.; Lingwood, C.A. Membrane cytosolic translocation of verotoxin A$_1$ subunit in target cells. *Microbiology* **2007**, *153*, 2700–2710. [CrossRef] [PubMed]

41. Zumbrun, S.D.; Hanson, L.; Sinclair, J.F.; Freedy, J.; Melton-Celsa, A.R.; Rodriguez-Canales, J.; Hanson, J.C.; O'Brien, A.D. Human intestinal tissue and cultured colonic cells contain globotriaosylceramide synthase mRNA and the alternate Shiga toxin receptor globotetraosylceramide. *Infect. Immun.* **2010**, *78*, 4488–4499. [CrossRef] [PubMed]

42. Lingwood, C.A.; Binnington, B.; Manis, A.; Branch, D.R. Globotriaosyl ceramide receptor function—where membrane structure and pathology intersect. *FEBS Lett.* **2010**, *584*, 1879–1886. [CrossRef] [PubMed]

43. Raa, H.; Grimmer, S.; Schwudke, D.; Bergan, J.; Walchli, S.; Skotland, T.; Shevchenko, A.; Sandvig, K. Glycosphingolipid requirements for endosome-to-Golgi transport of Shiga toxin. *Traffic* **2009**, *10*, 868–882. [CrossRef] [PubMed]

44. Kiarash, A.; Boyd, B.; Lingwood, C.A. Glycosphingolipid receptor function is modified by fatty acid content. Verotoxin 1 and verotoxin 2c preferentially recognize different globotriaosyl ceramide fatty acid homologues. *J. Biol. Chem.* **1994**, *269*, 11138–11146. [PubMed]

45. Pellizzari, A.; Pang, H.; Lingwood, C.A. Binding of verocytotoxin 1 to its receptor is influenced by differences in receptor fatty acid content. *Biochemistry* **1992**, *31*, 1363–1370. [CrossRef] [PubMed]

46. Sandvig, K.; Bergan, J.; Kavaliauskiene, S.; Skotland, T. Lipid requirements for entry of protein toxins into cells. *Prog. Lipid Res.* **2014**, *54*, 1–13. [CrossRef] [PubMed]

47. Kavaliauskiene, S.; Nymark, C.-M.; Bergan, J.; Simm, R.; Sylvanne, T.; Simolin, H.; Ekroos, K.; Skotland, T.; Sandvig, K. Cell density-induced changes in lipid composition and intracellular trafficking. *Cell. Mol. Life Sci.* **2013**, *71*, 1097–1116. [CrossRef] [PubMed]
48. Kavaliauskiene, S.; Torgersen, M.L.; Dyve Lingelem, A.B.; Klokk, T.I.; Lintonen, T.; Simolin, H.; Ekroos, K.; Skotland, T.; Sandvig, K. Cellular effects of fluorodeoxyglucose: Global changes in the lipidome and alteration in intracellular transport. *Oncotarget* **2016**, *7*, 79885–79900. [CrossRef] [PubMed]
49. Sandvig, K.; Ryd, M.; Garred, O.; Schweda, E.; Holm, P.K.; van Deurs, B. Retrograde transport from the Golgi complex to the ER of both Shiga toxin and the nontoxic Shiga B-fragment is regulated by butyric acid and cAMP. *J. Cell Biol.* **1994**, *126*, 53–64. [CrossRef] [PubMed]
50. Bergan, J.; Skotland, T.; Sylvänne, T.; Simolin, H.; Ekroos, K.; Sandvig, K. The ether lipid precursor hexadecylglycerol causes major changes in the lipidome of HEp-2 cells. *PLoS ONE* **2013**, *8*, e75904. [CrossRef] [PubMed]
51. Hughes, A.K.; Stricklett, P.K.; Schmid, D.; Kohan, D.E. Cytotoxic effect of Shiga toxin-1 on human glomerular epithelial cells. *Kidney Int.* **2000**, *57*, 2350–2359. [CrossRef] [PubMed]
52. Ramegowda, B.; Samuel, J.E.; Tesh, V.L. Interaction of Shiga toxins with human brain microvascular endothelial cells: Cytokines as sensitizing agents. *J. Infect. Dis.* **1999**, *180*, 1205–1213. [CrossRef] [PubMed]
53. Zumbrun, S.D.; Melton-Celsa, A.R.; O'Brien, A.D. When a healthy diet turns deadly. *Gut Microbes* **2014**, *5*, 40–43. [CrossRef] [PubMed]
54. Skotland, T.; Ekroos, K.; Kavaliauskiene, S.; Bergan, J.; Kauhanen, D.; Lintonen, T.; Sandvig, K. Determining the turnover of glycosphingolipid species by stable-isotope tracer lipidomics. *J. Mol. Biol.* **2016**, *428*, 4856–4866. [CrossRef] [PubMed]
55. Mahfoud, R.; Manis, A.; Lingwood, C.A. Fatty acid-dependent globotriaosyl ceramide receptor function in detergent resistant model membranes. *J. Lipid Res.* **2009**, *50*, 1744–1755. [CrossRef] [PubMed]
56. Lingwood, D.; Binnington, B.; Rog, T.; Vattulainen, I.; Grzybek, M.; Coskun, U.; Lingwood, C.A.; Simons, K. Cholesterol modulates glycolipid conformation and receptor activity. *Nat. Chem. Biol.* **2011**, *7*, 260–262. [CrossRef] [PubMed]
57. Mahfoud, R.; Manis, A.; Binnington, B.; Ackerley, C.; Lingwood, C.A. A major fraction of glycosphingolipids in model and cellular cholesterol-containing membranes is undetectable by their binding proteins. *J. Biol. Chem.* **2010**, *285*, 36049–36059. [CrossRef] [PubMed]
58. Boerlin, P.; McEwen, S.A.; Boerlin-Petzold, F.; Wilson, J.B.; Johnson, R.P.; Gyles, C.L. Associations between virulence factors of Shiga toxin-producing *Escherichia coli* and disease in humans. *J. Clin. Microbiol.* **1999**, *37*, 497–503. [PubMed]
59. Itoh, K.; Tezuka, T.; Inoue, K.; Tada, H.; Suzuki, T. Different binding property of verotoxin-1 and verotoxin-2 against their glycolipid receptor, globotriaosylceramide. *Tohoku J. Exp. Med.* **2001**, *195*, 237–243. [CrossRef] [PubMed]
60. Head, S.C.; Karmali, M.A.; Lingwood, C.A. Preparation of VT1 and VT2 hybrid toxins from their purified dissociated subunits. Evidence for B subunit modulation of a subunit function. *J. Biol. Chem.* **1991**, *266*, 3617–3621. [PubMed]
61. Rutjes, N.W.; Binnington, B.A.; Smith, C.R.; Maloney, M.D.; Lingwood, C.A. Differential tissue targeting and pathogenesis of verotoxins 1 and 2 in the mouse animal model. *Kidney Int.* **2002**, *62*, 832–845. [CrossRef] [PubMed]
62. Chark, D.; Nutikka, A.; Trusevych, N.; Kuzmina, J.; Lingwood, C. Differential carbohydrate epitope recognition of globotriaosyl ceramide by verotoxins and a monoclonal antibody. *Eur. J. Biochem.* **2004**, *271*, 405–417. [CrossRef] [PubMed]
63. Khan, F.; Proulx, F.; Lingwood, C.A. Detergent-resistant globotriaosyl ceramide may define verotoxin/glomeruli-restricted hemolytic uremic syndrome pathology. *Kidney Int.* **2009**, *75*, 1209–1216. [CrossRef] [PubMed]
64. Tam, P.; Mahfoud, R.; Nutikka, A.; Khine, A.A.; Binnington, B.; Paroutis, P.; Lingwood, C. Differential intracellular transport and binding of verotoxin 1 and verotoxin 2 to globotriaosylceramide-containing lipid assemblies. *J. Cell. Physiol.* **2008**, *216*, 750–763. [CrossRef] [PubMed]
65. Mukhopadhyay, S.; Redler, B.; Linstedt, A.D. Shiga toxin-binding site for host cell receptor GPP130 reveals unexpected divergence in toxin-trafficking mechanisms. *Mol. Biol. Cell* **2013**, *24*, 2311–2318. [CrossRef] [PubMed]

66. Selyunin, A.S.; Mukhopadhyay, S. A conserved structural motif mediates retrograde trafficking of Shiga toxin types 1 and 2. *Traffic* **2015**, *16*, 1270–1287. [CrossRef] [PubMed]

67. Ergonul, Z.; Clayton, F.; Fogo, A.B.; Kohan, D.E. Shigatoxin-1 binding and receptor expression in human kidneys do not change with age. *Pediatr. Nephrol.* **2003**, *18*, 246–253. [PubMed]

68. Lingwood, C.A. Verotoxin-binding in human renal sections. *Nephron* **1994**, *66*, 21–28. [CrossRef] [PubMed]

69. Boyd, B.; Lingwood, C. Verotoxin receptor glycolipid in human renal tissue. *Nephron* **1989**, *51*, 207–210. [CrossRef] [PubMed]

70. Ohmi, K.; Kiyokawa, N.; Takeda, T.; Fujimoto, J. Human microvascular endothelial cells are strongly sensitive to Shiga toxins. *Biochem. Biophys. Res. Commun.* **1998**, *251*, 137–141. [CrossRef] [PubMed]

71. Miyamoto, Y.; Iimura, M.; Kaper, J.B.; Torres, A.G.; Kagnoff, M.F. Role of Shiga toxin versus H7 flagellin in enterohaemorrhagic *Escherichia coli* signalling of human colon epithelium *in vivo*. *Cell. Microbiol.* **2006**, *8*, 869–879. [CrossRef] [PubMed]

72. Ren, J.; Utsunomiya, I.; Taguchi, K.; Ariga, T.; Tai, T.; Ihara, Y.; Miyatake, T. Localization of verotoxin receptors in nervous system. *Brain Res.* **1999**, *825*, 183–188. [CrossRef]

73. Cooling, L.L.; Walker, K.E.; Gille, T.; Koerner, T.A. Shiga toxin binds human platelets via globotriaosylceramide (P^k antigen) and a novel platelet glycosphingolipid. *Infect. Immun.* **1998**, *66*, 4355–4366. [PubMed]

74. Tao, R.V.; Sweeley, C.C.; Jamieson, G.A. Sphingolipid composition of human platelets. *J. Lipid Res.* **1973**, *14*, 16–25. [PubMed]

75. Steffensen, R.; Carlier, K.; Wiels, J.; Levery, S.B.; Stroud, M.; Cedergren, B.; Nilsson, S.B.; Bennett, E.P.; Jersild, C.; Clausen, H. Cloning and expression of the histo-blood group P^k UDP-galactose: Galβ1-4G1cβ1-Cer α1,4-galactosyltransferase. Molecular genetic basis of the p phenotype. *J. Biol. Chem.* **2000**, *275*, 16723–16729. [CrossRef] [PubMed]

76. Mangeney, M.; Richard, Y.; Coulaud, D.; Tursz, T.; Wiels, J. CD77: An antigen of germinal center B cells entering apoptosis. *Eur. J. Immunol.* **1991**, *21*, 1131–1140. [CrossRef] [PubMed]

77. Engedal, N.; Skotland, T.; Torgersen, M.L.; Sandvig, K. Shiga toxin and its use in targeted cancer therapy and imaging. *Microb. Biotechnol.* **2011**, *4*, 32–46. [CrossRef] [PubMed]

78. Okuda, T.; Tokuda, N.; Numata, S.; Ito, M.; Ohta, M.; Kawamura, K.; Wiels, J.; Urano, T.; Tajima, O.; Furukawa, K.; et al. Targeted disruption of Gb3/CD77 synthase gene resulted in the complete deletion of globo-series glycosphingolipids and loss of sensitivity to verotoxins. *J. Biol. Chem.* **2006**, *281*, 10230–10235. [CrossRef] [PubMed]

79. Lauvrak, S.U.; Walchli, S.; Iversen, T.G.; Slagsvold, H.H.; Torgersen, M.L.; Spilsberg, B.; Sandvig, K. Shiga toxin regulates its entry in a Syk-dependent manner. *Mol. Biol. Cell* **2006**, *17*, 1096–1109. [CrossRef] [PubMed]

80. Katagiri, Y.U.; Mori, T.; Nakajima, H.; Katagiri, C.; Taguchi, T.; Takeda, T.; Kiyokawa, N.; Fujimoto, J. Activation of Src family kinase yes induced by Shiga toxin binding to globotriaosyl ceramide (Gb3/CD77) in low density, detergent-insoluble microdomains. *J. Biol. Chem.* **1999**, *274*, 35278–35282. [CrossRef] [PubMed]

81. Mori, T.; Kiyokawa, N.; Katagiri, Y.U.; Taguchi, T.; Suzuki, T.; Sekino, T.; Sato, N.; Ohmi, K.; Nakajima, H.; Takeda, T.; et al. Globotriaosyl ceramide (CD77/Gb3) in the glycolipid-enriched membrane domain participates in B-cell receptor-mediated apoptosis by regulating lyn kinase activity in human B cells. *Exp. Hematol.* **2000**, *28*, 1260–1268. [CrossRef]

82. Torgersen, M.L.; Walchli, S.; Grimmer, S.; Skanland, S.S.; Sandvig, K. Protein kinase Cδ is activated by Shiga toxin and regulates its transport. *J. Biol. Chem.* **2007**, *282*, 16317–16328. [CrossRef] [PubMed]

83. Walchli, S.; Skanland, S.S.; Gregers, T.F.; Lauvrak, S.U.; Torgersen, M.L.; Ying, M.; Kuroda, S.; Maturana, A.; Sandvig, K. The mitogen-activated protein kinase p38 links Shiga toxin-dependent signaling and trafficking. *Mol. Biol. Cell* **2008**, *19*, 95–104. [CrossRef] [PubMed]

84. Klokk, T.I.; Kavaliauskiene, S.; Sandvig, K. Cross-linking of glycosphingolipids at the plasma membrane: consequences for intracellular signaling and traffic. *Cell. Mol. Life Sci.* **2016**, *73*, 1301–1316. [CrossRef] [PubMed]

85. Tcatchoff, L.; Andersson, S.; Utskarpen, A.; Klokk, T.I.; Skanland, S.S.; Pust, S.; Gerke, V.; Sandvig, K. Annexin A1 and A2: Roles in retrograde trafficking of Shiga toxin. *PLoS ONE* **2012**, *7*, e40429. [CrossRef] [PubMed]

86. Takenouchi, H.; Kiyokawa, N.; Taguchi, T.; Matsui, J.; Katagiri, Y.U.; Okita, H.; Okuda, K.; Fujimoto, J. Shiga toxin binding to globotriaosyl ceramide induces intracellular signals that mediate cytoskeleton remodeling in human renal carcinoma-derived cells. *J. Cell Sci.* **2004**, *117*, 3911–3922. [CrossRef] [PubMed]

87. Hehnly, H.; Longhini, K.M.; Chen, J.L.; Stamnes, M. Retrograde Shiga toxin trafficking is regulated by ARHGAP21 and Cdc42. *Mol. Biol. Cell* **2009**, *20*, 4303–4312. [CrossRef] [PubMed]

88. Hehnly, H.; Sheff, D.; Stamnes, M. Shiga toxin facilitates its retrograde transport by modifying microtubule dynamics. *Mol. Biol. Cell* **2006**, *17*, 4379–4389. [CrossRef] [PubMed]

89. Mallard, F.; Antony, C.; Tenza, D.; Salamero, J.; Goud, B.; Johannes, L. Direct pathway from early/recycling endosomes to the Golgi apparatus revealed through the study of Shiga toxin B-fragment transport. *J. Cell Biol.* **1998**, *143*, 973–990. [CrossRef] [PubMed]

90. Sandvig, K.; Olsnes, S.; Brown, J.E.; Petersen, O.W.; van Deurs, B. Endocytosis from coated pits of Shiga toxin: a glycolipid-binding protein from *Shigella dysenteriae* 1. *J. Cell Biol.* **1989**, *108*, 1331–1343. [CrossRef] [PubMed]

91. Sandvig, K.; Garred, O.; Prydz, K.; Kozlov, J.V.; Hansen, S.H.; van Deurs, B. Retrograde transport of endocytosed Shiga toxin to the endoplasmic reticulum. *Nature* **1992**, *358*, 510–512. [CrossRef] [PubMed]

92. Donta, S.T.; Tomicic, T.K.; Donohue-Rolfe, A. Inhibition of Shiga-like toxins by brefeldin A. *J. Infect. Dis.* **1995**, *171*, 721–724. [CrossRef] [PubMed]

93. Sandvig, K.; Prydz, K.; Ryd, M.; van Deurs, B. Endocytosis and intracellular transport of the glycolipid-binding ligand Shiga toxin in polarized MDCK cells. *J. Cell Biol.* **1991**, *113*, 553–562. [CrossRef] [PubMed]

94. Garred, O.; Dubinina, E.; Polesskaya, A.; Olsnes, S.; Kozlov, J.; Sandvig, K. Role of the disulfide bond in Shiga toxin A-chain for toxin entry into cells. *J. Biol. Chem.* **1997**, *272*, 11414–11419. [PubMed]

95. Garred, O.; Dubinina, E.; Holm, P.K.; Olsnes, S.; van Deurs, B.; Kozlov, J.V.; Sandvig, K. Role of processing and intracellular transport for optimal toxicity of Shiga toxin and toxin mutants. *Exp. Cell Res.* **1995**, *218*, 39–49. [CrossRef] [PubMed]

96. Johannes, L.; Romer, W. Shiga toxin—From cell biology to biomedical applications. *Nat. Rev. Microbiol.* **2010**, *8*, 105–116. [CrossRef] [PubMed]

97. Lee, M.S.; Koo, S.; Jeong, D.G.; Tesh, V.L. Shiga toxins as multi-functional proteins: Induction of host cellular stress responses, role in pathogenesis and therapeutic applications. *Toxins (Basel)* **2016**, *8*, 77. [CrossRef] [PubMed]

98. Dyve Lingelem, A.B.; Bergan, J.; Sandvig, K. Inhibitors of intravesicular acidification protect against Shiga toxin in a pH-independent manner. *Traffic* **2012**, *13*, 443–454. [CrossRef] [PubMed]

99. Kavaliauskiene, S.; Skotland, T.; Sylvanne, T.; Simolin, H.; Klokk, T.I.; Torgersen, M.L.; Lingelem, A.B.; Simm, R.; Ekroos, K.; Sandvig, K. Novel actions of 2-deoxy-d-glucose: protection against Shiga toxins and changes in cellular lipids. *Biochem. J.* **2015**, *470*, 23–37. [CrossRef] [PubMed]

100. Stechmann, B.; Bai, S.K.; Gobbo, E.; Lopez, R.; Merer, G.; Pinchard, S.; Panigai, L.; Tenza, D.; Raposo, G.; Beaumelle, B.; et al. Inhibition of retrograde transport protects mice from lethal ricin challenge. *Cell* **2010**, *141*, 231–242. [CrossRef] [PubMed]

101. Secher, T.; Shima, A.; Hinsinger, K.; Cintrat, J.C.; Johannes, L.; Barbier, J.; Gillet, D.; Oswald, E. Retrograde trafficking inhibitor of Shiga toxins reduces morbidity and mortality of mice infected with enterohemorrhagic *Esherichia coli*. *Antimicrob. Agents Chemother.* **2015**, *59*, 5010–5013. [CrossRef]

102. Tewari, R.; Jarvela, T.; Linstedt, A.D. Manganese induces oligomerization to promote down-regulation of the intracellular trafficking receptor used by Shiga toxin. *Mol. Biol. Cell* **2014**, *25*, 3049–3058. [CrossRef] [PubMed]

103. Mukhopadhyay, S.; Linstedt, A.D. Manganese blocks intracellular trafficking of Shiga toxin and protects against Shiga toxicosis. *Science* **2012**, *335*, 332–335. [CrossRef] [PubMed]

104. Gaston, M.A.; Pellino, C.A.; Weiss, A.A. Failure of manganese to protect from Shiga toxin. *PLoS ONE* **2013**, *8*, e69823. [CrossRef] [PubMed]

105. Sandvig, K.; Brown, J.E. Ionic requirements for entry of Shiga toxin from *Shigella dysenteriae* 1 into cells. *Infect. Immun.* **1987**, *55*, 298–303. [PubMed]

106. Silberstein, C.; Copeland, D.P.; Chiang, W.; Repetto, H.A.; Ibarra, C. A glucosylceramide synthase inhibitor prevents the cytotoxic effects of Shiga toxin-2 on human renal tubular epithelial cells. *J. Epithel. Biol. Pharmacol.* **2008**, *1*, 71–75. [CrossRef]

107. Silberstein, C.; Lucero, M.S.; Zotta, E.; Copeland, D.P.; Lingyun, L.; Repetto, H.A.; Ibarra, C. A glucosylceramide synthase inhibitor protects rats against the cytotoxic effects of Shiga toxin 2. *Pediatr. Res.* **2011**, *69*, 390–394. [CrossRef] [PubMed]

108. Amaral, M.M.; Sacerdoti, F.; Jancic, C.; Repetto, H.A.; Paton, A.W.; Paton, J.C.; Ibarra, C. Action of Shiga toxin type-2 and subtilase cytotoxin on human microvascular endothelial cells. *PLoS ONE* **2013**, *8*, e70431. [CrossRef] [PubMed]

109. Bergan, J.; Skotland, T.; Lingelem, A.B.; Simm, R.; Spilsberg, B.; Lindback, T.; Sylvanne, T.; Simolin, H.; Ekroos, K.; Sandvig, K. The ether lipid precursor hexadecylglycerol protects against Shiga toxins. *Cell. Mol. Life Sci.* **2014**, *71*, 4285–4300. [CrossRef] [PubMed]

110. Binnington, B.; Nguyen, L.; Kamani, M.; Hossain, D.; Marks, D.L.; Budani, M.; Lingwood, C.A. Inhibition of Rab prenylation by statins induces cellular glycosphingolipid remodeling. *Glycobiology* **2016**, *26*, 166–180. [CrossRef] [PubMed]

111. Becker, G.L.; Lu, Y.; Hardes, K.; Strehlow, B.; Levesque, C.; Lindberg, I.; Sandvig, K.; Bakowsky, U.; Day, R.; Garten, W.; et al. Highly potent inhibitors of proprotein convertase furin as potential drugs for treatment of infectious diseases. *J. Biol. Chem.* **2012**, *287*, 21992–22003. [CrossRef] [PubMed]

112. Hardes, K.; Becker, G.L.; Lu, Y.; Dahms, S.O.; Kohler, S.; Beyer, W.; Sandvig, K.; Yamamoto, H.; Lindberg, I.; Walz, L.; et al. Novel Furin inhibitors with potent anti-infectious activity. *ChemMedChem* **2015**, *10*, 1218–1231. [CrossRef] [PubMed]

113. Meshnick, S.R.; Dobson, M.J. The history of antimalarial drugs. In *Antimalarial Chemotherapy: Mechanisms of Action, Resistance, and New Directions in Drug Discovery*; Rosenthal, P.J., Ed.; Humana Press: Totowa, NJ, USA, 2001; pp. 15–25.

114. Kitchen, L.W.; Vaughn, D.W.; Skillman, D.R. Role of US military research programs in the development of US Food and Drug Administration—Approved antimalarial drugs. *Clin. Infect. Dis.* **2006**, *43*, 67–71. [CrossRef] [PubMed]

115. Al-Bari, M.A. Chloroquine analogues in drug discovery: New directions of uses, mechanisms of actions and toxic manifestations from malaria to multifarious diseases. *J. Antimicrob. Chemother.* **2015**, *70*, 1608–1621. [CrossRef] [PubMed]

116. Pascolo, S. Time to use a dose of chloroquine as an adjuvant to anti-cancer chemotherapies. *Eur. J. Pharmacol.* **2016**, *771*, 139–144. [CrossRef] [PubMed]

117. Keeling, D.J.; Herslof, M.; Ryberg, B.; Sjogren, S.; Solvell, L. Vacuolar H(+)-ATPases. Targets for drug discovery? *Ann. N. Y. Acad. Sci.* **1997**, *834*, 600–608. [CrossRef] [PubMed]

118. Parra, K.J. Vacuolar ATPase: A model proton pump for antifungal drug discovery. In *Antimicrobial Drug Discovery: Emerging Strategies*; Tegos, G., Mylonakis, E., Eds.; CAB International: Accra, Ghana, 2012; pp. 89–100.

119. Moreau, D.; Kumar, P.; Wang, S.C.; Chaumet, A.; Chew, S.Y.; Chevalley, H.; Bard, F. Genome-wide RNAi screens identify genes required for Ricin and PE intoxications. *Dev. Cell* **2011**, *21*, 231–244. [CrossRef] [PubMed]

120. Yu, M.; Haslam, D.B. Shiga toxin is transported from the endoplasmic reticulum following interaction with the luminal chaperone HEDJ/ERDj3. *Infect. Immun.* **2005**, *73*, 2524–2532. [CrossRef] [PubMed]

121. Rodriguez-Menchaca, A.A.; Navarro-Polanco, R.A.; Ferrer-Villada, T.; Rupp, J.; Sachse, F.B.; Tristani-Firouzi, M.; Sanchez-Chapula, J.A. The molecular basis of chloroquine block of the inward rectifier Kir2.1 channel. *Proc. Natl. Acad. Sci. USA* **2008**, *105*, 1364–1368. [CrossRef] [PubMed]

122. Orlik, F.; Schiffler, B.; Benz, R. Anthrax toxin protective antigen: inhibition of channel function by chloroquine and related compounds and study of binding kinetics using the current noise analysis. *Biophys J.* **2005**, *88*, 1715–1724. [CrossRef] [PubMed]

123. Bachmeyer, C.; Benz, R.; Barth, H.; Aktories, K.; Gilbert, M.; Popoff, M.R. Interaction of *Clostridium botulinum* C2 toxin with lipid bilayer membranes and Vero cells: inhibition of channel function by chloroquine and related compounds *in vitro* and intoxification *in vivo*. *FASEB J.* **2001**, *15*, 1658–1660. [CrossRef] [PubMed]

124. Browning, D.J. Pharmacology of chloroquine and hydroxychloroquine. In *Hydroxychloroquine and Chloroquine Retinopathy*; Springer: New York, NY, USA, 2014; pp. 35–63.

125. Molina, D.K. Postmortem hydroxychloroquine concentrations in nontoxic cases. *Am. J. Forensic Med. Pathol.* **2012**, *33*, 41–42. [CrossRef] [PubMed]

126. Brown, J. Effects of 2-deoxyglucose on carbohydrate metablism: review of the literature and studies in the rat. *Metabolism* **1962**, *11*, 1098–1112. [PubMed]

127. Wick, A.N.; Drury, D.R.; Nakada, H.I.; Wolfe, J.B. Localization of the primary metabolic block produced by 2-deoxyglucose. *J. Biol. Chem.* **1957**, *224*, 963–969. [PubMed]

128. Cramer, F.B.; Woodward, G.E. 2-Desoxy-D-glucose as an antagonist of glucose in yeast fermentation. *J. Frankl. Inst.* **1952**, *253*, 354–360. [CrossRef]

129. Sols, A.; Crane, R.K. Substrate specificity of brain hexokinase. *J. Biol. Chem.* **1954**, *210*, 581–595. [PubMed]

130. Chen, W.; Gueron, M. The inhibition of bovine heart hexokinase by 2-deoxy-D-glucose-6-phosphate: characterization by ^{31}P NMR and metabolic implications. *Biochimie* **1992**, *74*, 867–873. [CrossRef]

131. Datema, R.; Schwarz, R.T. Interference with glycosylation of glycoproteins. Inhibition of formation of lipid-linked oligosaccharides *in vivo*. *Biochem. J.* **1979**, *184*, 113–123. [CrossRef] [PubMed]

132. Desselle, A.; Chaumette, T.; Gaugler, M.H.; Cochonneau, D.; Fleurence, J.; Dubois, N.; Hulin, P.; Aubry, J.; Birkle, S.; Paris, F. Anti-Gb3 monoclonal antibody inhibits angiogenesis and tumor development. *PLoS ONE* **2012**, *7*, e45423. [CrossRef] [PubMed]

133. Watowich, S.S.; Morimoto, R.I. Complex regulation of heat shock- and glucose-responsive genes in human cells. *Mol. Cell. Biol.* **1988**, *8*, 393–405. [CrossRef] [PubMed]

134. Shinjo, S.; Mizotani, Y.; Tashiro, E.; Imoto, M. Comparative analysis of the expression patterns of UPR-target genes caused by UPR-inducing compounds. *Biosci. Biotechnol. Biochem.* **2013**, *77*, 729–735. [CrossRef] [PubMed]

135. Okuda, T.; Furukawa, K.; Nakayama, K.I. A novel, promoter-based, target-specific assay identifies 2-deoxy-D-glucose as an inhibitor of globotriaosylceramide biosynthesis. *FEBS J.* **2009**, *276*, 5191–5202. [CrossRef] [PubMed]

136. Kurtoglu, M.; Maher, J.C.; Lampidis, T.J. Differential toxic mechanisms of 2-deoxy-D-glucose versus 2-fluorodeoxy-D-glucose in hypoxic and normoxic tumor cells. *Antioxid. Redox Signal.* **2007**, *9*, 1383–1390. [CrossRef] [PubMed]

137. Lampidis, T.J.; Kurtoglu, M.; Maher, J.C.; Liu, H.; Krishan, A.; Sheft, V.; Szymanski, S.; Fokt, I.; Rudnicki, W.R.; Ginalski, K.; et al. Efficacy of 2-halogen substituted D-glucose analogs in blocking glycolysis and killing "hypoxic tumor cells". *Cancer Chemother. Pharmacol.* **2006**, *58*, 725–734. [CrossRef] [PubMed]

138. Datema, R.; Schwarz, R.T.; Jankowski, A.W. Fluoroglucose-inhibition of protein glycosylation *in vivo*. Inhibition of mannose and glucose incorporation into lipid-linked oligosaccharides. *Eur. J. Biochem.* **1980**, *109*, 331–341. [CrossRef] [PubMed]

139. Schmidt, M.F.; Biely, P.; Kratky, Z.; Schwarz, R.T. Metabolism of 2-deoxy-2-fluoro-D-[^{3}H]glucose and 2-deoxy-2-fluoro-D-[^{3}H]mannose in yeast and chick-embryo cells. *Eur. J. Biochem.* **1978**, *87*, 55–68. [CrossRef] [PubMed]

140. Hoh, C.K. Clinical use of FDG-PET. *Nucl. Med. Biol.* **2007**, *34*, 737–742. [CrossRef] [PubMed]

141. Kelloff, G.J.; Hoffman, J.M.; Johnson, B.; Scher, H.I.; Siegel, B.A.; Cheng, E.Y.; Cheson, B.D.; O'Shaughnessy, J.; Guyton, K.Z.; Mankoff, D.A.; et al. Progress and promise of FDG-PET imaging for cancer patient management and oncologic drug development. *Clin. Cancer. Res.* **2005**, *11*, 2785–2808. [CrossRef] [PubMed]

142. Xi, H.; Barredo, J.C.; Merchan, J.R.; Lampidis, T.J. Endoplasmic reticulum stress induced by 2-deoxyglucose but not glucose starvation activates AMPK through CAMKKβ leading to autophagy. *Biochem. Pharmacol.* **2013**, *85*, 1463–1477. [CrossRef] [PubMed]

143. Xi, H.; Kurtoglu, M.; Liu, H.; Wangpaichitr, M.; You, M.; Liu, X.; Savaraj, N.; Lampidis, T.J. 2-Deoxy-D-glucose activates autophagy via endoplasmic reticulum stress rather than ATP depletion. *Cancer Chemother. Pharmacol.* **2011**, *67*, 899–910. [CrossRef] [PubMed]

144. Biswas, C.; Ostrovsky, O.; Makarewich, C.A.; Wanderling, S.; Gidalevitz, T.; Argon, Y. The peptide-binding activity of GRP94 is regulated by calcium. *Biochem. J.* **2007**, *405*, 233–241. [CrossRef] [PubMed]

145. Hendershot, L.M. The ER function BiP is a master regulator of ER function. *Mt. Sinai J. Med.* **2004**, *71*, 289–297. [PubMed]

146. Raez, L.E.; Papadopoulos, K.; Ricart, A.D.; Chiorean, E.G.; Dipaola, R.S.; Stein, M.N.; Rocha Lima, C.M.; Schlesselman, J.J.; Tolba, K.; Langmuir, V.K.; et al. A phase I dose-escalation trial of 2-deoxy-D-glucose alone or combined with docetaxel in patients with advanced solid tumors. *Cancer Chemother. Pharmacol.* **2013**, *71*, 523–530. [CrossRef] [PubMed]

147. Saenz, J.B.; Doggett, T.A.; Haslam, D.B. Identification and characterization of small molecules that inhibit intracellular toxin transport. *Infect. Immun.* **2007**, *75*, 4552–4561. [CrossRef] [PubMed]

148. Park, J.G.; Kahn, J.N.; Tumer, N.E.; Pang, Y.P. Chemical structure of Retro-2, a compound that protects cells against ribosome-inactivating proteins. *Sci. Rep.* **2012**, *2*, 631. [CrossRef] [PubMed]

149. Noel, R.; Gupta, N.; Pons, V.; Goudet, A.; Garcia-Castillo, M.D.; Michau, A.; Martinez, J.; Buisson, D.A.; Johannes, L.; Gillet, D.; et al. *N*-methyldihydroquinazolinone derivatives of Retro-2 with enhanced efficacy against Shiga toxin. *J. Med. Chem.* **2013**, *56*, 3404–3413. [CrossRef] [PubMed]

150. Carney, D.W.; Nelson, C.D.; Ferris, B.D.; Stevens, J.P.; Lipovsky, A.; Kazakov, T.; DiMaio, D.; Atwood, W.J.; Sello, J.K. Structural optimization of a retrograde trafficking inhibitor that protects cells from infections by human polyoma- and papillomaviruses. *Bioorg. Med. Chem.* **2014**, *22*, 4836–4847. [CrossRef] [PubMed]

151. Gupta, N.; Pons, V.; Noel, R.; Buisson, D.A.; Michau, A.; Johannes, L.; Gillet, D.; Barbier, J.; Cintrat, J.C. (*S*)-*N*-Methyldihydroquinazolinones are the active enantiomers of Retro-2 derived compounds against toxins. *ACS Med. Chem. Lett.* **2014**, *5*, 94–97. [CrossRef] [PubMed]

152. Frank, C.; Werber, D.; Cramer, J.P.; Askar, M.; Faber, M.; an der Heiden, M.; Bernard, H.; Fruth, A.; Prager, R.; Spode, A.; et al. Epidemic profile of Shiga-toxin-producing *Escherichia coli* O104:H4 outbreak in Germany. *N. Engl. J. Med.* **2011**, *365*, 1771–1780. [CrossRef] [PubMed]

153. Bielaszewska, M.; Mellmann, A.; Zhang, W.; Kock, R.; Fruth, A.; Bauwens, A.; Peters, G.; Karch, H. Characterisation of the *Escherichia coli* strain associated with an outbreak of haemolytic uraemic syndrome in Germany, 2011: a microbiological study. *Lancet Infect. Dis.* **2011**, *11*, 671–676. [CrossRef]

154. Gupta, N.; Noel, R.; Goudet, A.; Hinsinger, K.; Michau, A.; Pons, V.; Abdelkafi, H.; Secher, T.; Shima, A.; Shtanko, O.; et al. Inhibitors of retrograde trafficking active against ricin and Shiga toxins also protect cells from several viruses, *Leishmania* and Chlamydiales. *Chem. Biol. Interact.* **2016**. [CrossRef] [PubMed]

155. Aschner, M.; Erikson, K.M.; Herrero Hernandez, E.; Tjalkens, R. Manganese and its role in Parkinson's disease: from transport to neuropathology. *Neuromol. Med.* **2009**, *11*, 252–266. [CrossRef] [PubMed]

156. Racette, B.A.; Aschner, M.; Guilarte, T.R.; Dydak, U.; Criswell, S.R.; Zheng, W. Pathophysiology of manganese-associated neurotoxicity. *Neurotoxicology* **2012**, *33*, 881–886. [CrossRef] [PubMed]

157. Vunnam, R.R.; Radin, N.S. Analogs of ceramide that inhibit glucocerebroside synthetase in mouse brain. *Chem. Phys. Lipids* **1980**, *26*, 265–278. [CrossRef]

158. Jmoudiak, M.; Futerman, A.H. Gaucher disease: pathological mechanisms and modern management. *Br. J. Haematol.* **2005**, *129*, 178–188. [CrossRef] [PubMed]

159. Barbour, S.; Edidin, M.; Felding-Habermann, B.; Taylor-Norton, J.; Radin, N.S.; Fenderson, B.A. Glycolipid depletion using a ceramide analogue (PDMP) alters growth, adhesion, and membrane lipid organization in human A431 cells. *J. Cell Physiol.* **1992**, *150*, 610–619. [CrossRef] [PubMed]

160. Arab, S.; Lingwood, C.A. Intracellular targeting of the endoplasmic reticulum/nuclear envelope by retrograde transport may determine cell hypersensitivity to verotoxin via globotriaosyl ceramide fatty acid isoform traffic. *J. Cell. Physiol.* **1998**, *177*, 646–660. [CrossRef]

161. Sandvig, K.; Garred, O.; van Helvoort, A.; van Meer, G.; van Deurs, B. Importance of glycolipid synthesis for butyric acid-induced sensitization to Shiga toxin and intracellular sorting of toxin in A431 cells. *Mol. Biol. Cell* **1996**, *7*, 1391–1404. [CrossRef] [PubMed]

162. Lee, L.; Abe, A.; Shayman, J.A. Improved inhibitors of glucosylceramide synthase. *J. Biol. Chem.* **1999**, *274*, 14662–14669. [CrossRef] [PubMed]

163. Basu, M.; Dastgheib, S.; Girzadas, M.A.; O'Donnell, P.H.; Westervelt, C.W.; Li, Z.; Inokuchi, J.; Basu, S. Hydrophobic nature of mammalian ceramide glycanases: purified from rabbit and rat mammary tissues. *Acta Biochim. Pol.* **1998**, *45*, 327–342. [PubMed]

164. Abe, A.; Gregory, S.; Lee, L.; Killen, P.D.; Brady, R.O.; Kulkarni, A.; Shayman, J.A. Reduction of globotriaosylceramide in Fabry disease mice by substrate deprivation. *J. Clin. Invest.* **2000**, *105*, 1563–1571. [CrossRef] [PubMed]

165. Poole, R.M. Eliglustat: first global approval. *Drugs* **2014**, *74*, 1829–1836. [CrossRef] [PubMed]

166. Belmatoug, N.; Di Rocco, M.; Fraga, C.; Giraldo, P.; Hughes, D.; Lukina, E.; Maison-Blanche, P.; Merkel, M.; Niederau, C.; Plckinger, U.; et al. Management and monitoring recommendations for the use of eliglustat in adults with type 1 Gaucher disease in Europe. *Eur. J. Intern. Med.* **2016**. [CrossRef] [PubMed]

167. Lukina, E.; Watman, N.; Arreguin, E.A.; Banikazemi, M.; Dragosky, M.; Iastrebner, M.; Rosenbaum, H.; Phillips, M.; Pastores, G.M.; Rosenthal, D.I.; et al. A phase 2 study of eliglustat tartrate (Genz-112638), an oral substrate reduction therapy for Gaucher disease type 1. *Blood* **2010**, *116*, 893–899. [CrossRef] [PubMed]

168. Das, A.K.; Holmes, R.D.; Wilson, G.N.; Hajra, A.K. Dietary ether lipid incorporation into tissue plasmalogens of humans and rodents. *Lipids* **1992**, *27*, 401–405. [CrossRef] [PubMed]

169. Iannitti, T.; Palmieri, B. An update on the therapeutic role of alkylglycerols. *Mar. Drugs* **2010**, *8*, 2267–2300. [CrossRef] [PubMed]

170. Deniau, A.L.; Mosset, P.; Pedrono, F.; Mitre, R.; Le Bot, D.; Legrand, A.B. Multiple beneficial health effects of natural alkylglycerols from shark liver oil. *Mar. Drugs* **2010**, *8*, 2175–2184. [CrossRef] [PubMed]

171. Matusewicz, L.; Meissner, J.; Toporkiewicz, M.; Sikorski, A.F. The effect of statins on cancer cells—Review. *Tumour Biol.* **2015**, *36*, 4889–4904. [CrossRef] [PubMed]

172. Correale, M.; Abruzzese, S.; Greco, C.A.; Concilio, M.; Di Biase, M.; Brunetti, N.D. Statins in heart failure. *Curr. Vasc. Pharmacol.* **2012**, *69*, 232. [CrossRef]

173. Paliani, U.; Ricci, S. The role of statins in stroke. *Intern. Emerg. Med.* **2012**, *7*, 305–311. [CrossRef] [PubMed]

174. US Preventive Services Task Force; Bibbins-Domingo, K.; Grossman, D.C.; Curry, S.J.; Davidson, K.W.; Epling, J.W.; Garcia, F.A.; Gillman, M.W.; Kemper, A.R.; Krist, A.H.; et al. Statin use for the primary prevention of cardiovascular disease in adults: US preventive services task force recommendation statement. *JAMA* **2016**, *316*, 1997–2007. [PubMed]

175. Oda, T.; Wu, H.C. Effect of lovastatin on the cytotoxicity of ricin, modeccin, *Pseudomonas* toxin, and diphtheria toxin in brefeldin A-sensitive and -resistant cell lines. *Exp. Cell Res.* **1994**, *212*, 329–337. [CrossRef] [PubMed]

176. Gomes, A.Q.; Ali, B.R.; Ramalho, J.S.; Godfrey, R.F.; Barral, D.C.; Hume, A.N.; Seabra, M.C. Membrane targeting of Rab GTPases is influenced by the prenylation motif. *Mol. Biol. Cell* **2003**, *14*, 1882–1899. [CrossRef] [PubMed]

177. Pfeffer, S.; Aivazian, D. Targeting Rab GTPases to distinct membrane compartments. *Nat. Rev. Mol. Cell Biol.* **2004**, *5*, 886–896. [CrossRef] [PubMed]

178. Holstein, S.A.; Knapp, H.R.; Clamon, G.H.; Murry, D.J.; Hohl, R.J. Pharmacodynamic effects of high dose lovastatin in subjects with advanced malignancies. *Cancer Chemother. Pharmacol.* **2006**, *57*, 155–164. [CrossRef] [PubMed]

179. Schafer, W.; Stroh, A.; Berghofer, S.; Seiler, J.; Vey, M.; Kruse, M.L.; Kern, H.F.; Klenk, H.D.; Garten, W. Two independent targeting signals in the cytoplasmic domain determine trans-Golgi network localization and endosomal trafficking of the proprotein convertase furin. *EMBO J.* **1995**, *14*, 2424–2435. [PubMed]

180. Plaimauer, B.; Mohr, G.; Wernhart, W.; Himmelspach, M.; Dorner, F.; Schlokat, U. 'Shed' furin: Mapping of the cleavage determinants and identification of its C-terminus. *Biochem. J.* **2001**, *354*, 689–695. [CrossRef] [PubMed]

181. Thomas, G. Furin at the cutting edge: from protein traffic to embryogenesis and disease. *Nat. Rev. Mol. Cell Biol.* **2002**, *3*, 753–766. [CrossRef] [PubMed]

182. Seidah, N.G.; Prat, A. The biology and therapeutic targeting of the proprotein convertases. *Nat. Rev. Drug Discov.* **2012**, *11*, 367–383. [CrossRef] [PubMed]

183. Couture, F.; D'Anjou, F.; Day, R. On the cutting edge of proprotein convertase pharmacology: from molecular concepts to clinical applications. *Biomol. Concepts* **2011**, *2*, 421–438. [CrossRef] [PubMed]

184. De Cicco, R.L.; Bassi, D.E.; Benavides, F.; Conti, C.J.; Klein-Szanto, A.J. Inhibition of proprotein convertases: approaches to block squamous carcinoma development and progression. *Mol. Carcinog.* **2007**, *46*, 654–659. [CrossRef] [PubMed]

185. Sarac, M.S.; Cameron, A.; Lindberg, I. The furin inhibitor hexa-D-arginine blocks the activation of *Pseudomonas aeruginosa* exotoxin A in vivo. *Infect. Immun.* **2002**, *70*, 7136–7139. [CrossRef] [PubMed]

186. Sarac, M.S.; Peinado, J.R.; Leppla, S.H.; Lindberg, I. Protection against anthrax toxemia by hexa-D-arginine in vitro and in vivo. *Infect. Immun.* **2004**, *72*, 602–605. [CrossRef] [PubMed]

187. Basak, A. Inhibitors of proprotein convertases. *J. Mol. Med. (Berl.)* **2005**, *83*, 844–855. [CrossRef] [PubMed]

188. Shayman, J.A. The design and clinical development of inhibitors of glycosphingolipid synthesis: Will invention be the mother of necessity? *Trans. Am. Clin. Climatol. Assoc.* **2013**, *124*, 46–60. [PubMed]

189. Liu, Y.Y.; Hill, R.A.; Li, Y.T. Ceramide glycosylation catalyzed by glucosylceramide synthase and cancer drug resistance. *Adv. Cancer Res.* **2013**, *117*, 59–89. [PubMed]

Article

Probiotic Microorganisms Inhibit Epithelial Cell Internalization of Botulinum Neurotoxin Serotype A

Tina I. Lam †, Christina C. Tam †, Larry H. Stanker and Luisa W. Cheng *

Foodborne Toxin Detection and Prevention Research Unit, Western Regional Research Center, Agricultural Research Service, United States Department of Agriculture, 800 Buchanan Street, Albany, CA 94710, USA; tinaiunsanlam@gmail.com (T.I.L.); christina.tam@ars.usda.gov (C.C.T.); larry.stanker@ars.usda.gov (L.H.S.)

* Correspondence: luisa.cheng@ars.usda.gov; Tel.: +1-510-559-6337; Fax: +1-510-559-5880

† These authors contributed equally to this work.

Academic Editor: Holger Barth
Received: 30 September 2016; Accepted: 13 December 2016; Published: 16 December 2016

Abstract: Botulinum neurotoxins (BoNTs) are some of the most poisonous natural toxins known to man and are threats to public health and safety. Previous work from our laboratory showed that both BoNT serotype A complex and holotoxin can bind and transit through the intestinal epithelia to disseminate in the blood. The timing of BoNT/A toxin internalization was shown to be comparable in both the Caco-2 in vitro cell culture and in the oral mouse intoxication models. Probiotic microorganisms have been extensively studied for their beneficial effects in not only maintaining the normal gut mucosa but also protection from allergens, pathogens, and toxins. In this study, we evaluate whether probiotic microorganisms will block BoNT/A uptake in the in vitro cell culture system using Caco-2 cells. Several probiotics tested (*Saccharomyces boulardii, Lactobacillus acidophilus, Lactobacillus rhamnosus* LGG, and *Lactobacillus reuteri*) blocked BoNT/A uptake in a dose-dependent manner whereas a non-probiotic strain of *Escherichia coli* did not. We also showed that inhibition of BoNT/A uptake was not due to the degradation of BoNT/A nor by sequestration of toxin via binding to probiotics. These results show for the first time that probiotic treatment can inhibit BoNT/A binding and internalization in vitro and may lead to the development of new therapies.

Keywords: botulinum toxin; foodborne toxins; probiotic bacteria; toxin absorption

1. Introduction

Botulinum neurotoxins are produced by the ubiquitous, gram-positive, anaerobic spore-forming *Clostridium* species and are the causative agent of botulism [1,2]. There are at least seven, possibly eight, different serotypes of BoNTs (A–H) of which A, B, E, and F are known causes of botulism in humans [3–7]. BoNTs are highly poisonous to humans with a parenteral lethal dosage of 0.1–1 ng/kg and an oral dose of 1 µg/kg. They are classified by the Centers for Disease Control and Prevention (CDC) as among the highest threats for bioterrorism (Tier 1 Category A agents). Additionally, BoNTs remain a public health and safety threat in the form of foodborne, wound, and infant botulism. Due to its mortality and morbidity, there is a significant economic burden associated with the long-term management of intoxication.

BoNTs are A-B dimeric toxins synthesized as ~150 kDa holotoxin with a heavy chain ~100 kDa linked by a disulfide bond to the light chain ~50 kDa. There are three functional domains: a receptor binding domain (H_C), translocation domain (H_N), and a catalytic domain (LC) [7]. The preferential target cells for BoNTs are the peripheral cholinergic neurons. Binding of H_C to carbohydrate and protein receptors on the presynaptic membrane results in BoNT endocytosis [8–10]. Internalization of BoNTs leads to H_N pore formation in the endosomal membrane resulting in the translocation of the catalytic domain LC into the cytosol [11–14]. The catalytic domain LC is a zinc-dependent endopeptidase that cleaves proteins associated with intracellular vesicular transport such as SNAP-25 (synaptosome-associated protein of 25 kDa), VAMP (vesicle-associated membrane protein), or syntaxin [15–17]. Due to the cleavage of these mediators of intracellular transport, exocytosis of the neurotransmitter acetylcholine from neurons is inhibited causing flaccid muscle paralysis. In foodborne illnesses caused by BoNTs, toxins must be able to survive initially in the lumen of the gastrointestinal tract, then bind and translocate through the intestinal epithelium to reach the bloodstream. Previous work from our laboratory showed that the BoNT/A complex, comprised of the combination of holotoxin with neurotoxin-associated proteins (NAPs), binds and transits through the intestinal epithelia to disseminate in the blood faster than BoNT/A holotoxin alone [18]. Therefore, understanding the mechanism(s) in which BoNTs bind to and breach this epithelial barrier is of great scientific interest because of the potential development of new therapeutics to inhibit this required first step of oral intoxication.

The gastrointestinal tract (GI) has evolved as one of the largest barriers to segregate the extracellular milieu from mammalian cells. Colonization of the gastrointestinal tract by a variety of commensal bacteria aid in not only the digestion and absorption of nutrients but also the development and regulation of the mucosal immune system [19]. There are anywhere between 10^{10} to 10^{12} colony-forming units per gram of intestinal content in the colon and 60% of all fecal matter mass in humans is due to bacteria [20]. The colonization of the GI tract with microbes carries with it the risk of infection and inflammation if the barrier between the microorganisms and hosts is damaged. The intestinal epithelium acts as a physical and biochemical barrier to not only commensal and pathogenic bacteria but also to all other luminal contents including other injurious matter such as toxins. Specialized intestinal epithelial cells (IECs) are able to sense and respond to these stimuli with appropriate responses such as increasing their barrier function to activation of anti-pathogenic immune mechanisms [19].

Probiotics, as defined by the World Health Organization (WHO), are live microorganisms that provide health benefits to hosts when ingested in adequate amounts. They have been shown to have potential significant therapeutic value for a range of diseases such as *H. pylori* infection, irritable bowel syndrome, and inflammatory bowel disease (ulcerative colitis and Crohn's disease) as well as boosting the immune system of healthy individuals [21–26]. The most common probiotic strains used are *Lactobacillus*, *Bifidobacteria*, and the yeast strain *Saccharomyces cerevisiae* var *boulardii* (SB). Lactic acid bacteria and bifidobacteria have been shown to remove heavy metals [27], cyanotoxins [28], and mycotoxin from in vitro aqueous solutions [29,30]. The probiotic effects seen are both strain and species dependent indicating that combinations of different strains and species may need to be tailored to the specific issue at hand rather than having one "universal" probiotic therapy. Though some beneficial effects of probiotics have been shown in both in vivo and in vitro studies, the exact mechanism(s) that is responsible for these beneficial effects remains to be fully elucidated. The mechanisms that have been attributed to probiotics are: (a) maintenance of the gut epithelial barrier, (b) competitive exclusion of pathogenic organisms, (c) secretion of antimicrobial products, and (d) regulation of the mucosal immune system in favor of the hosts.

Since probiotics have been shown to block pathogen internalization as well as remove heavy metals and some toxins, we wondered if probiotics may block entry and subsequent internalization of BoNT/A in an in vitro cell based assay system using Caco-2 cells.

2. Results

2.1. The Effect of Pre-Treatment with Saccharomyces Boulardii on BoNT/A Uptake in Caco-2 Cells

Previous work in our laboratory established two in vitro Caco-2 cell models to test the entry and subsequent internalization of BoNT/A holotoxin and BoNT/A complex (AC) [18]. This study showed that BoNT/A entry and internalization was enhanced by the presence of neurotoxin-associated proteins in the BoNT/A complex. Significant internalization of toxin complex was achieved by 4 h post-intoxication whereas holotoxin was slightly delayed. Since some probiotics have been shown to be important for inhibition of pathogens as well as toxin binding to host mammalian cells, we wondered if pre-treatment with the probiotic yeast strain *Saccharomyces boulardii* (SB) would have a negative effect on BoNT/A binding and internalization in Caco-2 cells.

We chose a simple Caco-2 cell model to study the effect of probiotics on BoNT/A entry. This in vitro model was shown to mirror results found through the in vitro polarized Caco-2 epithelial cell and in the mouse oral intoxication model [18]. Caco-2 cells were either pre-treated with media (control) or SB low (10^4 CFU) or high (10^8 CFU) concentrations for 30 min at 37 °C before removal of non-adherent SB and subsequent washing with $1\times$ HBSS three times. Cells were then incubated with 50 ng/mL of BoNT/A toxin complex at 37 °C for 4 h. At the end of this incubation, cells were washed and fixed with 4% paraformaldehyde. Immunostaining was performed to detect BoNT/A toxin using a polyclonal rabbit anti-BoNT/A antibody with detection using a goat anti-rabbit-IgG-Alexa-488. Additionally, these coverslips were stained with Rhodamine-Phalloidin to delineate the actin cytoskeleton in mammalian cells. DAPI was used to stain nuclear DNA. Images were obtained throughout the depth of the cells to measure the internalization of BoNT/A. Mean fluorescence intensities were measured throughout the depth of each field of cells (Z stack) containing the same area. Mean intensity multiplied by area for each Z section was calculated and the sum of the total fluorescence was defined as BoNT/A signal indicating cellular uptake. The mean of the BoNT/A signal was calculated and the statistical significance was determined for each condition.

A statistically significant cellular uptake of BoNT/A after 4 h incubation of BoNT/A toxin complex as compared to the control treated with media alone (BoNT/A vs. control, $p = 0.0004$) is seen in Figure 1A. BoNT/A present in Caco-2 cells was visualized with bright green fluorescence while the actin cytoskeleton was stained in red and DAPI stained the cellular nuclei blue. Signal intensity for the different conditions were quantified in Figure 1B. Figure 1 shows that pre-treatment with SB 30 min prior to toxin addition has a negative effect on BoNT/A uptake. The addition of 10^8 CFU of SB reduces the BoNT/A fluorescence signal (decrease in green fluorescence) even more than 10^4 CFU, indicating that there is a dose-dependent effect on internalization and that it is statistically significant (BoNT/A vs. BoNT/A + Low SB, $p = 0.0181$; BoNT/A vs. BoNT/A + High SB, $p = 0.0013$).

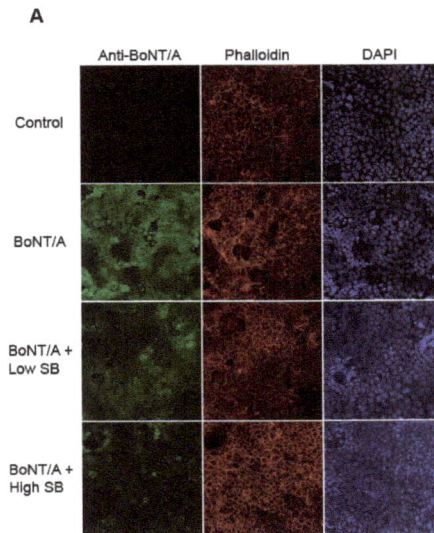

Figure 1. *Cont.*

B

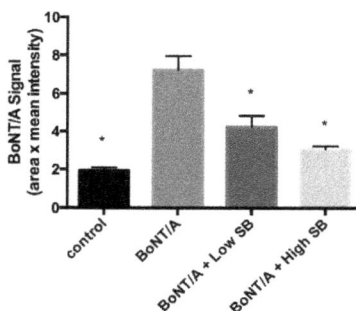

Figure 1. Internalization of BoNT/A into Caco-2 cells is significantly reduced in a dose-dependent manner by pre-treatment with the probiotic *Saccharomyces boulardii*. (**A**) Caco-2 cells were treated with media (control) or BoNT/A complex for 4 h at 37 °C. Some Caco-2 cells were either pre-treated with *Saccharomyces boulardii* (SB) for 30 min at 37 °C at either high SB (10^8 CFU) or low SB (10^4 CFU) before the addition of BoNT/A. Cells were fixed and stained with Alexa-488 labeled antibodies to BoNT/A, DAPI (nuclear), and Rhodamine-Phalloidin (actin cytoskeleton). Representative images at 40× magnification are shown; (**B**) The cellular uptake of BoNT/A was quantified by determining the mean fluorescence of three randomly chosen optical fields from each of four coverslips per experiment using ImageJ software. Values represent means of four independent experiments ± SEM. Statistical significance was determined using two-way ANOVA followed by the Tukey-Kramer test where multiple groups that are compared with p-values < 0.05 are taken to indicate significant differences between groups (*).

2.2. The Effect of Pre-Treatment with Escherichia coli MG1655 on BoNT/A Uptake in Caco-2 Cells

Since we see a significant reduction on the internalization of BoNT/A into Caco-2 cells with SB in a dose-dependent manner, one can argue that this decrease in cellular uptake may be inherently due to non-specific interactions between any microorganism given in sufficient quantities. To assess whether this hypothesis was true, we asked whether a non-probiotic strain of *Escherichia coli* K12 MG1655 (EC) can reduce the uptake of BoNT/A in Caco-2 cells. As seen in Figure 2A,B, we show that EC does not have any statistically significant effect on BoNT/A uptake either at a low dose (10^4 CFU) or a high dose (10^8 CFU) unlike SB (Figure 1A,B). However, a statistical difference was seen between the control media alone compared to the sample treated with BoNT/A complex (control vs. BoNT/A, $p < 0.0001$). These results support the hypothesis that probiotics can have a beneficial effect on blocking the internalization of the foodborne toxin botulinum neurotoxin serotype A.

2.3. The Effect of Pre-Treatment with Lactobacillus Acidophilus, Lactobacillus rhamnosus LGG, and Lactobacillus Reuteri on BoNT/A Uptake in Caco-2 Cells

We showed in Figure 1 that the probiotic yeast strain SB has a beneficial effect on decreasing the uptake of BoNT/A in our in vitro cell culture system. We wanted to ask if other probiotic strains such as Lactobacilli would have a similar effect because some Lactobacilli strains such as *Lactobacillus acidophilus* (LA), *Lactobacillus rhamnosus* (LGG), and *Lactobacillus reuteri* (Lr) have been shown to have beneficial effects on host physiology against pathogen and toxin injuries.

Lactobacilli treatment severely reduces BoNT/A uptake into Caco-2 cells (Figure 3A,B). This severe inhibition of BoNT/A internalization by LA, LGG, and Lr is more dramatic than with SB (Figure 3A,B vs. Figure 1A,B). Additionally, the dose-dependent BoNT/A uptake decrease seen with SB does not seem to happen with LA, LGG, or Lr. A low dose (10^4 CFU) of LA, LGG, and Lr is sufficient to almost completely block BoNT/A internalization to the levels seen in control cells which have not seen toxin (BoNT/A vs. LA Low, $p = 0.0018$; BoNT/A vs. LGG Low, $p = 0.0019$; BoNT/A vs. Lr Low, $p = 0.0007$). These results suggest that there may be differences in the ability of various probiotic strains to efficiently block some foodborne toxins such as BoNT/A.

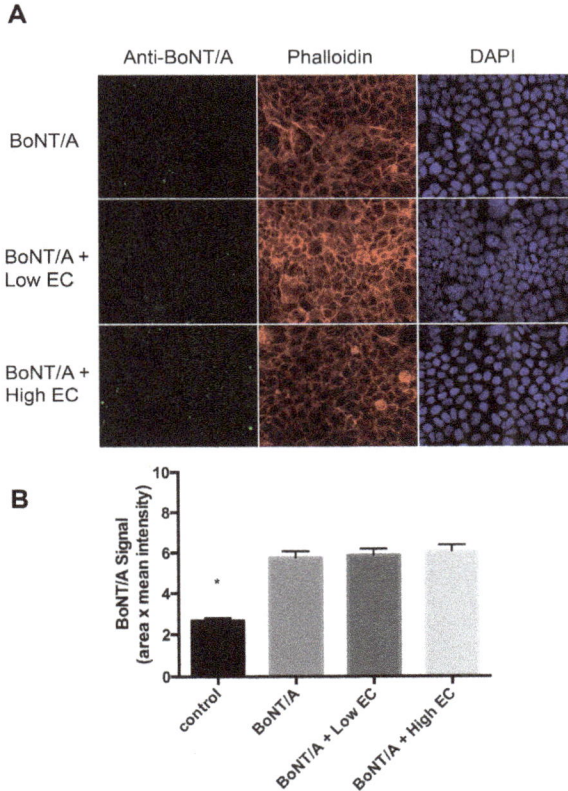

Figure 2. Pre-treatment with *Escherichia coli* MG1655 does not affect the internalization of BoNT/A into Caco-2. (**A**) Caco-2 cells were treated with media (control) or BoNT/A complex for 4 h at 37 °C. Some Caco-2 cells were either pre-treated with *Escherichia coli* (EC) for 30 min at 37 °C at either high EC (10^8 CFU) or low EC (10^4 CFU) before the addition of toxin. Cells were fixed and stained with Alexa-488 labeled antibodies to BoNT/A, DAPI (nuclear), and Rhodamine-Phalloidin (actin cytoskeleton). Representative images at 40× magnification are shown; (**B**) The cellular uptake of BoNT/A was quantified by determining the mean fluorescence of three randomly chosen optical fields from each of three coverslips per experiment acquired using a Zeiss Axio Observer.Z1 with Apotome.2 and analyzed with Zeiss Zen Pro 2012 software. Values represent means of four independent experiments ± SEM. Statistical significance was determined using two-way ANOVA followed by the Tukey-Kramer test where multiple groups that are compared with p-values < 0.05 are taken to indicate significant differences between groups (*). There is no statistical significance with pretreatment with EC.

A

B

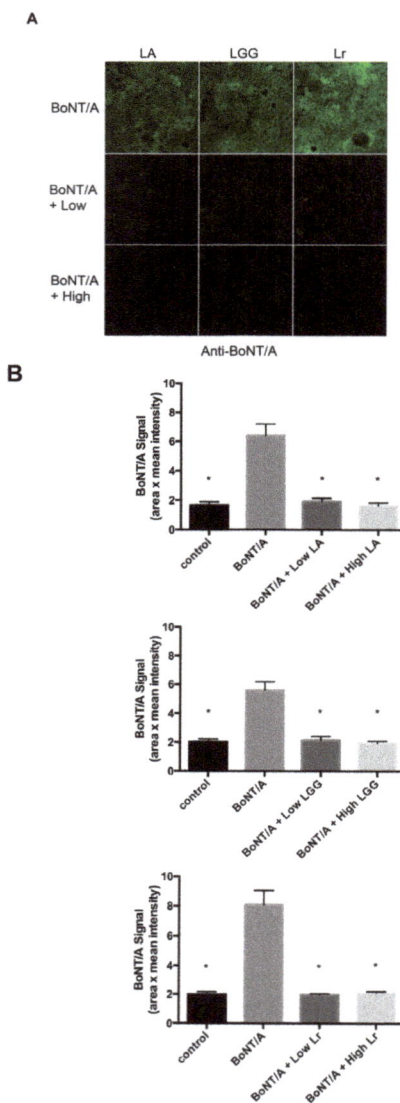

Figure 3. Pre-treatment with the probiotic *Lactobacillus acidophilus*, *Lactobacillus rhamnosus*, and *Lactobacillus reuteri* blocks internalization of BoNT/A into Caco-2 cells. (**A**) Caco-2 cells were treated with media (control) or with BoNT/A for 4 h at 37 °C. Some Caco-2 cells were either pre-treated with *Lactobacillus acidophilus* (LA), *Lactobacillus rhamnosus* (LGG), *or Lactobacillus reuteri* (Lr) for 30 min at 37 °C at either high (10^8 CFU) or low (10^4 CFU) before addition of BoNT/A complex. Cells were fixed and stained with Alexa-488 labeled antibodies to BoNT/A, DAPI (nuclear), and Rhodamine-Phalloidin (actin cytoskeleton). Representative images at 40× magnification showing BoNT/A fluorescence are shown; (**B**) The cellular uptake of BoNT/A was quantified by determining the mean fluorescence of three randomly chosen optical fields from each of the four coverslips per strain per experiment using ImageJ software. Values represent means of four independent experiments ± SEM. Statistical significance was determined using two-way ANOVA followed by the Tukey-Kramer test where multiple groups are compared with p-values < 0.05 are taken to indicate significant differences between groups (*).

2.4. Evaluation of the Mechanism Used by Probiotic Strains to Block BoNT/A Internalization

We have shown that treatment with SB, LA, LGG, and Lr strains prior to the addition of BoNT/A complex reduced the internalization of BoNT/A toxin in a colonic adenocarcinoma cell model. However, the mechanism(s) used by probiotics to block BoNT/A internalization is still unclear. One potential mechanism of action would be for these probiotics to secrete proteases that could degrade BoNT/A and hence there would be less BoNT/A to bind to and be internalized into the cells. Another mode of action would be for non-specific binding of BoNT/A to the probiotics themselves (i.e., cell walls) thus sequestering BoNT/A from its cellular receptors and not allowing for binding and subsequent internalization. A third potential mechanism is for the probiotics to compete for binding with BoNT/A to its cellular receptors, thus blocking binding and internalization.

We sought to answer this important question using a co-precipitation assay with the results detected using an antibody that recognizes BoNT/A in Western blots. BoNT/A (2 µg/mL) was added to an aliquot of washed overnight bacterial cultures (EC, SB, LA, LGG, and Lr) in $1\times$ HBSS and then incubated for 4 h at 37 °C. The samples were centrifuged to fractionate BoNT/A into soluble supernatant (unbound) and insoluble pellet (bound). Protein samples from each fraction were TCA-precipitated, solubilized in protein sample buffer, and prepared for SDS-PAGE electrophoresis. Proteins were transferred onto PVDF membrane and immunoblotting was performed. A rabbit polyclonal anti-BoNT/A was used to bind to BoNT/A and detection was enabled by the addition of a goat anti-rabbit IgG conjugated to horseradish-peroxidase. Chemiluminescent substrate was added and signal was detected using an AlphaImager. Densitometry was used to detect signal intensity using the FluorChemSP.

In Figure 4B, full length BoNT/A is predominately detected in Western blots at the expected size of ~150 kDa whether treated with EC, SB, LA, LGG, or Lr. This result suggests that the mechanism of action with any of the probiotics to decrease BoNT/A internalization is not due to the degradation of BoNT/A by secreted probiotic proteases (Figure 4A,B). As expected, treatment with EC, a non-probiotic strain that does not decrease BoNT/A uptake in cells, fractionates the BoNT/A mainly in the soluble supernatant unbound fraction ~83% with ~17% found in the insoluble bound pellet. BoNT/A is found exclusively in the soluble supernatant fraction and not bound to SB in the pellet (EC pellet vs. SB pellet, $p = 0.0272$). LA, LGG, and Lr were found to be even more efficient than SB in inhibiting BoNT/A uptake and all three show BoNT/A is present predominately in the soluble supernatant fraction similar to EC. These results suggest that probiotics themselves are not non-specifically sequestering toxins from mammalian cells. Thus, the most likely mechanism of action is due to competition between the probiotics and BoNT/A for the same cellular receptors either directly or via steric-hindrance.

A

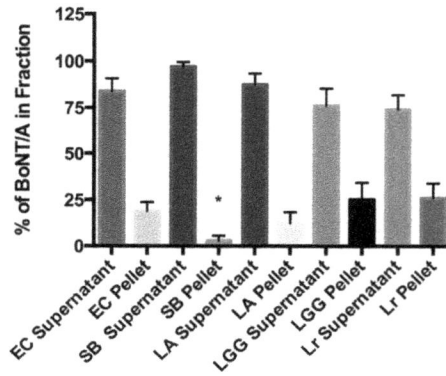

B

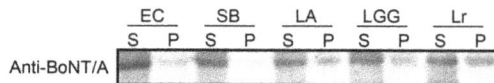

Figure 4. Decreased cellular uptake of BoNT/A complex is not due to the proteolytic degradation of holotoxin nor binding of toxin to probiotics. BoNT/A was added to either *Escherichia coli* MG1655, or probiotics and incubated for 4 h at 37 °C. Soluble supernatant (S) and insoluble pellet (P) fractions were precipitated with trichloracetic acid (TCA). Precipitates were solubilized with sample loading buffer and loaded onto 10% Bis-Tris NuPage gels. Gels were transferred onto PVDF membranes and incubated with primary polyclonal antibody to BoNT/A (Metabiologics) and secondary goat anti-rabbit-HRP. Western blot was developed using Pierce SuperSignal ECL substrate. (**A**) Mean percent signal of BoNT/A in each fraction was quantified from four independent experiments ± SEM using FluorChem SP (Alpha Innotech); (**B**) Representative Western depicting the presence of full length BoNT/A. Statistical significance was determined by a two-tailed unpaired Student's *t*-test, (*) $p < 0.05$.

3. Discussion

Botulinum neurotoxins, with their potential contamination of food, are bioterror threats as well as public health hazards. Consumption of botulinum neurotoxins from food sources leads to muscle paralysis and/or death for humans. There is a significant economic burden due to botulinum intoxication because of the need for long term supportive care and intensive hospitalization associated with this disease. Therefore, studies to elucidate the initial entry and internalization process of the toxin in the gut is of critical importance because of the potential development of new therapies to proactively block intoxication or in ameliorating the function of the toxin after ingestion.

BoNTs also cause infant botulism, which is usually associated with the ingestion of foods contaminated with *Clostridia* spores. Ingestion and subsequent germination of these spores into viable neurotoxigenic bacteria that are able to colonize the infant gastrointestinal system due to the lack of a robust gut microbiota to outcompete *Clostridia* [1,31]. BoNTs after ingestion or in situ production from bacteria must be able to survive in the lumen of the gastrointestinal tract and then traverse the intestinal epithelium from the apical to the basolateral side to reach its target cells.

The mechanism(s) as to how this occurs has been a major focus in the field. Previous work showed that the majority of toxin absorption occurs in the mouse upper small intestine [7,32]. One model for the transit of BoNTs suggests that the holotoxin itself can transcytose through the intestinal epithelium [33–35]. A second model for the BoNT absorption from the epithelium implicates the hemagglutin proteins (HA) in this process by binding cell surface receptors on the apical side, transcytosis, and potential disruption of the epithelial barrier at the basolateral side to allow for paracellular transport of the toxins in certain situations [33,36]. Our lab has shown in in vitro and in vivo intoxication models that neurotoxin-accessory proteins enhanced the rate of entry of BoNT/A in comparison to holotoxin alone, and entry was localized first to intestinal villi and subsequently to the intestinal crypts [18].

The major defensive mechanism of the gut is thus the intestinal barrier, which maintains epithelial integrity, and protects the host from the environment. In defense of this barrier, there are also the mucous layer, antimicrobial peptides, secretory IgA, and the epithelial junction adhesion complex [37]. Disruption of this barrier allows for bacteria and food antigens to reach the submucosa, which can induce an inflammatory response potentially leading to the intestinal disorders such as inflammatory bowel disease [38,39]. Probiotic treatment has been shown to have many beneficial effects including: (a) therapeutic treatment for human diseases, (b) inhibition of growth and toxin production for pathogens, and (c) extraction of heavy metals and toxins (aflatoxin B1) from solution.

Studies have suggested that probiotics enhance the expression of genes involved in tight junction signaling as a possible mechanism to reinforce the integrity of the intestinal epithelium [40]. An example of this is that Lactobacilli treatment in a T84 cell barrier model modulates several genes such as E-cadherin and β-catenin that affect adherence cell junctions. Lactobacilli treatment of intestinal cells also differentially regulates the phosphorylation of adherence junction proteins and the abundance of protein kinase C (PKC) isoforms, such as PKCδ, thereby positively reinforcing epithelial barrier function [41]. Not only is the epithelial barrier reinforced before damage, work with the probiotic *Escherichia coli* Nissle 1917 strain (EcN1917) suggests that it can initiate repair of the mucosal barrier after damage by enteropathogenic *E. coli* in T84 and Caco-2 cells by enhancing the expression and redistribution of tight junction proteins of the zonula occludens (ZO-2) and PKC [42,43]. Similar repair mechanisms have been reported with treatment with *Lactobacillus casei* DN-114001 [44] and VSL3 (a pre- and probiotics mixture) [45].

Studies have also shown that probiotics are able to modify their environment to make it more hostile to their potential competitors. The production of antimicrobial substances such as lactic and acetic acid is one example of this modification. *Lactobacillus* cocultivation with *E. coli* O157:H7 in broth culture produced organic acids which lead to a decrease in both pH and stx_{2A} expression [46].

We have shown that pre-treatment with the yeast strain *Saccharomyces boulardii* significantly decreased BoNT/A binding and internalization in Caco-2 cells after 4 h in a dose-dependent and specific manner whereas the control non-probiotic strain *E. coli* did not (Figure 1A,B vs. Figure 2A,B). Treatment with *Lactobacillus acidophilus*, *Lactobacillus rhamnosus* LGG, and *Lactobacillus reuteri* demonstrated an even greater protective effect than *Saccharomyces boulardii* by almost completely abolishing BoNT/A binding and internalization at the lower 10^4 CFU dose (Figure 3A,B vs. Figure 1A,B). We have also tested the inhibition of BoNT/A binding using different commercial probiotic supplements, most showed inhibitory effects (data not shown). These results suggest that, consistent with other probiotic studies, the beneficial effects of probiotics are strain- and species-dependent [47,48]. The probiotic *E. coli* strain Nissle 1917 was however not available in the U.S. for use in comparison testing at the time of study. Further research using probiotic *E. coli* or comparable strains are needed to elucidate the mechanism of inhibition. Future studies are also needed to determine the right formulation or combination of probiotic strains for optimal toxin entry inhibition and the specific mechanisms of toxin entry inhibition.

What role do probiotic organisms play in the defense of the intestinal epithelium against toxic invader and block BoNT/A entry? One hypothesis would be that the toxin itself could be degraded by the probiotics via secretion of proteases, thus rendering the toxin unable to bind its cellular receptors. An alternative theory would be that the probiotics would non-specifically bind BoNT/A itself due to some constituent of their cell walls, thus titrating the toxin from the host cells. A third potential mechanism would competitive inhibition between the probiotics and BoNT/A for binding to the host cell either through direct exclusion of BoNT/A from binding to the host cell receptors or indirect exclusion of BoNT/A due to steric hindrance from probiotic binding to the cell membrane. The adhesive properties due to the interactions between surface proteins and mucins may be utilized by some probiotic strains as an antagonistic mechanism against gastrointestinal pathogens as well as some toxins. Since binding to the mucous layer and intestinal cells is required for entry and colonization by many pathogens and toxins, mechanisms that will inhibit this first required step are critical for disease prevention. To prevent enteric infections, approaches such as (a) the development of synthetic oligosaccharide-based anti-infectives such as Synsorb (inert silica particles-linked to synthetic oligosaccharides) have been developed against the following: $Stx1/2-Gb_3$, $Stx2e-Gb_4$, $Ctx-GM1$, $LT-GM1$, epsilon toxin-GM2, TcdA-Lewis X and Lewis Y, botulinum neurotoxin- GD1a, GT1b, *E. coli* K88 ad fimbriae-nLc4, and *E. coli* P pili- Gb_3 and Gb_4) and (b) recombinant receptor mimics against STEC [49].

There are a variety of mechanisms used by bacterial species to exclude or reduce the growth of another species such as creation of a hostile environment, blocking available receptor sites, production and secretion of antimicrobial products and specific metabolites, and competitive depletion of essential nutrients [50]. Lactobacilli and bifidobacteria have been shown to inhibit a broad range of pathogens including *E. coli*, *Salmonella*, *Helicobacter pylori*, *Listeria monocytogenes*, and *Rotavirus* [21,51–57]. It has been shown that some lactobacilli and bifidobacteria compete for binding to host cell receptors because they share the same carbohydrate-binding specificities with some enteropathogens [58–60]. *Lactobacillus rhamnosus* has been shown to inhibit the internalization of enterohemorrhagic *E. coli* (EHEC) [61]. Generally, the ability of probiotic strains to inhibit pathogen attachment relies on steric hindrance of enterocyte pathogen receptors [62].

In Figure 4A,B, westerns blots indicate the presence of full-length BoNT/A in the presence of all strains including the probiotics. This suggests that degradation of BoNT/A is not the mechanism used by the probiotics strains to interfere with BoNT/A entry. Densitometry analysis indicates that BoNT/A remains mostly soluble in the supernatant rather than sedimenting with the bacterial/yeast insoluble pellet (Figure 4A). Thus, the non-specific binding of probiotic bacteria to cell wall constituents therefore blocking the accessibility of BoNT/A receptors could be a likely mechanism that needs further investigation.

Our results suggest that the BoNT/A internalization could be blocked by the probiotics strains of SB and some lactobacilli species. For the first time, we show a potential beneficial role that some probiotics may have in blocking the development and/or limiting the effects of human botulism. These results may lead to the development of new therapeutics for food-borne but especially relevant in regards to infant botulism. Development of a probiotic "cocktail" to initially seed the undeveloped infant gut to establish an environment of beneficial gut microbiota may protect against both *Clostridia* colonization and/or toxin absorption.

4. Conclusions

We show that probiotics may be beneficial in preventing the binding and internalization of botulinum neurotoxin serotype A to mammalian cells. The data suggests that the mechanism involved in this process is competitive inhibition between the probiotic strains and BoNT/A for host cell membrane receptors rather than degradation of BoNT/A or non-specific binding of toxin to the probiotics themselves.

5. Materials and Methods

5.1. Materials

Dulbecco's modified Eagle's medium (DMEM) (containing 4.5 g/L D-glucose and GlutaMAX), penicillin and streptomycin (100×), fetal bovine serum (FBS), TrypLE Select, Hanks' balanced salt solution (HBSS), and phosphate buffer solution (PBS) ($10\times$) were purchased from Life Technologies (Carlsbad, CA, USA). The human colon carcinoma cell lines (Caco-2 cells, ATCC, Manassas, VA, USA) were grown in DMEM. Cells after 50–70 passages were used in the uptake study. *L. rhamnosus* LGG (ATCC 53103), *L. reuteri* (ATCC 23272), *L. acidophilus* (ATCC 4356), and *Saccharomyces boulardii* (ATCC MYA-796) were obtained from the American Type Culture Collection (ATCC, Manassas, VA, USA). All other chemicals and reagents used were obtained from Sigma-Aldrich (St. Louis, MO, USA). *Escherichia coli* K12 MG1655 (EC) was obtained from Dr. Lisa Gorski at the Western Regional Research Center.

5.2. Growth of Yeast and Bacterial Cultures

Escherichia coli K12 MG1655 (EC), *L. rhamnosus* (LGG), *L. reuteri* (Lr), *L. acidophilus* (LA), and *Saccharomyces cerevisiae* var *boulardii* (SB) were grown overnight in growth broth (LB, YPD, and MRS), washed two times with phosphate buffered saline (PBS) and resuspended in Hank's balanced salt solution (HBSS) at either 10^4 or 10^8 CFU/mL. The initial concentration was determined by spectrophotometry at 600 nm and the numbers of bacteria were verified by pour-plate assay using (LB, YPD, and MRS) agar and standard serial dilution techniques.

5.3. Caco-2 Culture and Pre-Treatment with Probiotic Cultures

Human colonic carcinoma Caco-2 cells were grown on acid-washed 25 mm glass coverslips incubated in DMEM containing 10% FBS, $1\times$ nonessential amino acid (NAA), and 1X penicillin and streptomycin at 37 °C in a 90% humidity and 5% CO_2 incubator (Sanyo, Osaka, Japan). Caco-2 cells were seeded at a density of 1.2×10^5 cells/well. The media was changed every two days. After five days, the cell monolayers were observed by optical microscopy (Leica Microsystems, Buffalo Grove, IL, USA) to ensure that the cells reached about 90% confluence. On the day of experiment, Caco-2 cell monolayers were washed with 1X HBSS two times and pretreated either with 10^4 or 10^8 EC, LGG, Lr, LA, or SC prepared in HBSS for 30 min. Non-adherent bacteria were removed from the culture with three washes of HBSS. After pretreatment, glass coverslips containing bacteria-bound Caco-2 monolayers were treated with BoNT/A complex, 50 ng·mL^{-1} (56 pM) in HBSS for 4 h. After incubation, the coverslips were washed three times with 1X PBS and fixed in 4% paraformaldehyde (PFA, Affymetrix, Santa Clara, CA, USA) for 10 min and rinsed with 1X PBS before immunofluorescence staining.

5.4. Immunofluorescence Staining

Fixed glass coverslips containing Caco-2 cells were rinsed twice with PBS and permeabilized with 1% triton X-100 in PBS for 30 min. Cells were incubated with blocking solution (2% goat serum, 0.2% Triton X-100, and 0.1% Bovine Serum Albumin) for 1 h, and then incubated with 1:250 blocking buffer diluted solutions of a polyclonal rabbit antibody against BoNT/A (2 mg·mL^{-1} of stock) and Rhodamine-Phalloidin (actin stain, Molecular Probes; Life Technologies). After washing three times, cells were incubated with Alexa Fluor 488 to rabbit IgG (1:500 dilution; Life Technologies) and mounted onto glass slides using the hard-set DAPI mounting medium (Vector Laboratories). Fluorescence signals from Z stacks representing the top to the bottom of optical fields were visualized with either the Leica Microsystems confocal microscope (Leica TCS SP5) or with Zeiss Axio Observer.Z1 microscope with Apotome.2 with the appropriate filters set at $40\times$ magnification. Negative controls were prepared by omitting the primary antibodies.

5.5. Western Blotting

Western blotting was performed to evaluate the probiotic mechanism of action with the BoNT/A neurotoxin. *E. coli*, *L. rhamnosus*, *L. reuteri*, *L. acidophilus*, and *Saccharomyces boulardii* were grown overnight in growth media (LB, YPD, and MRS). One mL of bacteria was washed three times with 1X HBSS and resuspended in 1 mL of HBSS. Aliquots of 50 µL bacteria mixture were taken from each strain and put in 1.5 mL Eppendorf Lo-bind tubes. Each tube containing bacteria was treated with 2 µL of

1 mg/mL BoNT/A complex for 4 h at 37 °C in a 90% humidity and 5% CO_2 incubator. After incubation, the tubes were centrifuged at 2000 × g for 5 min and the supernatant and pellet were separated. Protein samples were TCA precipitated and separated by sodium dodecyl sulfate (SDS)-polyacrylamide gel electrophoresis (PAGE) with NuPAGE 10% Bis-Tris gels (Invitrogen) followed by Western blotting. The resolved proteins were transferred to a PVDF membrane (Immobilon). The membrane was blocked in 5% milk-Tris-buffered saline-0.05% Tween 20 buffer then probed with polyclonal rabbit anti-BoNT/A antibody (2 mg·mL^{-1} of stock) diluted to 1:2000 with blocking solution followed by secondary antibody (Horseradish peroxidase (HRP)-conjugated; 1:2000). The blot was incubated in Pierce ECL Western Blotting Substrate solution (Thermo Scientific). Protein bands from peroxidase activities to chemiluminescent substrates were developed and detected using the FluorChem SP AlphaImager (Alpha Innotech, San Leandro, CA, USA). Molecular weight standards were purchased from Invitrogen. Densitometry was performed using the FluorChem analysis software. Percent of BoNT/A signal from the soluble and pellet was quantified from four independent experiments and plotted using GraphPad Prism 6.

5.6. Statistics

For the cell culture studies, n = number of independent experiments, each independent experiment contained triplicate culture wells with one coverslip per each study condition. BoNT/A signal, representing the total fluorescence intensity in the cells calculated from area multiplied by mean intensity, was quantified from at least 30 optical fields (Z stacks) taken from four independent experiments. For quantification, the mean fluorescence was measured in at least three randomly selected non-overlapping 40× fields with each containing approximately 100–150 Caco-2 cells. All data were expressed as mean ± standard error of the mean (SEM) and assessed using two-way ANOVA followed by the Tukey-Kramer test where multiple groups are compared with p values < 0.05 are taken to indicate significant differences between groups. Western data was assembled from four independent experiments and statistical significance was determined by two-tailed unpaired Student's t-test.

Acknowledgments: This work was funded by the United States Department of Agriculture, Agricultural Research Service, National Program project NP108, CRIS 5325-42000-049-00D. Larry H. Stanker was also funded by interagency agreement IAA#40768.

Author Contributions: L.W.C., T.I.L., and C.C.T. conceived and designed the experiments; T.I.L. and C.C.T. performed the experiments; L.W.C., T.I.L., and C.C.T. analyzed the data; L.H.S. contributed reagents; C.C.T., L.H.S., and L.W.C. wrote the paper.

Conflicts of Interest: The authors declare no conflict of interest.

Abbreviations

The following abbreviations are used in this manuscript:

BoNTs	Botulinum neurotoxins
NAPs	neurotoxin-associated proteins
CFU	colony forming unit
PBS	phosphate buffered saline

References

1. Arnon, S.S.; Schechter, R.; Inglesby, T.V.; Henderson, D.A.; Bartlett, J.G.; Ascher, M.S.; Eitzen, E.; Fine, A.D.; Hauer, J.; Layton, M.; et al. Botulinum toxin as a biological weapon: Medical and public health management. *J. Am. Med. Assoc.* **2001**, *285*, 1059–1070. [CrossRef]
2. Center for Disease Control and Prevention. *2015 Annual Report of the Federal Select Agent Program*; Center for Disease Control and Prevention: Atlanta, GA, USA, 2016.
3. Barash, J.R.; Arnon, S.S. A novel strain of *Clostridium botulinum* that produces type B and type H botulinum toxins. *J. Infect. Dis.* **2014**, *209*, 183–191. [CrossRef] [PubMed]

4. Tighe, A.P.; Schiavo, G. Botulinum neurotoxins: Mechanism of action. *Toxicon* **2013**, *67*, 87–93. [CrossRef] [PubMed]

5. Rummel, A. The long journey of botulinum neurotoxins into the synapse. *Toxicon* **2015**, *107*, 9–24. [CrossRef] [PubMed]

6. Hill, K.K.; Xie, G.; Foley, B.T.; Smith, T.J. Genetic diversity within the botulinum neurotoxin-producing bacteria and their neurotoxins. *Toxicon* **2015**, *107*, 2–8. [CrossRef] [PubMed]

7. Rossetto, O.; Pirazzini, M.; Montecucco, C. Botulinum neurotoxins: Genetic, structural and mechanistic insights. *Nat. Rev. Microbiol.* **2014**, *12*, 535–549. [CrossRef] [PubMed]

8. Dolly, J.O.; Black, J.; Williams, R.S.; Melling, J. Acceptors for botulinum neurotoxin reside on motor nerve terminals and mediate its internalization. *Nature* **1984**, *307*, 457–460. [CrossRef] [PubMed]

9. Dong, M.; Yeh, F.; Tepp, W.H.; Dean, C.; Johnson, E.A.; Janz, R.; Chapman, E.R. SV2 is the protein receptor for botulinum neurotoxin A. *Science* **2006**, *312*, 592–596. [CrossRef] [PubMed]

10. Montecucco, C.; Tonello, F.; Zanotti, G. Stop the killer: How to inhibit the anthrax lethal factor metalloprotease. *Trends Biochem. Sci.* **2004**, *29*, 282–285. [CrossRef] [PubMed]

11. Simpson, L.L. Identification of the major steps in botulinum toxin action. *Annu. Rev. Pharmacol. Toxicol.* **2004**, *44*, 167–193. [CrossRef] [PubMed]

12. Mahrhold, S.; Rummel, A.; Bigalke, H.; Davletov, B.; Binz, T. The synaptic vesicle protein 2c mediates the uptake of botulinum neurotoxin A into phrenic nerves. *FEBS Lett.* **2006**, *580*, 2011–2014. [CrossRef] [PubMed]

13. Blaustein, R.O.; Germann, W.J.; Finkelstein, A.; DasGupta, B.R. The N-terminal half of the heavy chain of botulinum type A neurotoxin forms channels in planar phospholipid bilayers. *FEBS Lett.* **1987**, *226*, 115–120. [CrossRef]

14. Fischer, A.; Montal, M. Single molecule detection of intermediates during botulinum neurotoxin translocation across membranes. *Proc. Natl. Acad. Sci. USA* **2007**, *104*, 10447–10452. [CrossRef] [PubMed]

15. Schiavo, G.; Poulain, B.; Benfenati, F.; DasGupta, B.R.; Montecucco, C. Novel targets and catalytic activities of bacterial protein toxins. *Trends Microbiol.* **1993**, *1*, 170–174. [CrossRef]

16. Montecucco, C.; Schiavo, G. Mechanism of action of tetanus and botulinum neurotoxins. *Mol. Microbiol.* **1994**, *13*, 1–8. [CrossRef] [PubMed]

17. Montecucco, C.; Papini, E.; Schiavo, G. Bacterial protein toxins and cell vesicle trafficking. *Experientia* **1996**, *52*, 1026–1032. [PubMed]

18. Lam, T.I.; Stanker, L.H.; Lee, K.; Jin, R.; Cheng, L.W. Translocation of botulinum neurotoxin serotype A and associated proteins across the intestinal epithelia. *Cell. Microbiol.* **2015**, *17*, 1133–1143. [CrossRef] [PubMed]

19. Peterson, L.W.; Artis, D. Intestinal epithelial cells: Regulators of barrier function and immune homeostasis. *Nat. Rev. Immunol.* **2014**, *14*, 141–153. [PubMed]

20. Stephen, A.M.; Cummings, J.H. The microbial contribution to human faecal mass. *J. Med. Microbiol.* **1980**, *13*, 45–56. [CrossRef] [PubMed]

21. Myllyluoma, E.; Veijola, L.; Ahlroos, T.; Tynkkynen, S.; Kankuri, E.; Vapaatalo, H.; Rautelin, H.; Korpela, R. Probiotic supplementation improves tolerance to helicobacter pylori eradication therapy—A placebo-controlled, double-blind randomized pilot study. *Aliment. Pharmacol. Ther.* **2005**, *21*, 1263–1272. [CrossRef] [PubMed]

22. Kajander, K.; Hatakka, K.; Poussa, T.; Farkkila, M.; Korpela, R. A probiotic mixture alleviates symptoms in irritable bowel syndrome patients: A controlled 6-month intervention. *Aliment. Pharmacol. Ther.* **2005**, *22*, 387–394. [CrossRef] [PubMed]

23. Olivares, M.; Diaz-Ropero, M.A.; Gomez, N.; Lara-Villoslada, F.; Sierra, S.; Maldonado, J.A.; Martin, R.; Lopez-Huertas, E.; Rodriguez, J.M.; Xaus, J. Oral administration of two probiotic strains, *Lactobacillus gasseri* CECT5714 and *Lactobacillus coryniformis* CECT5711, enhances the intestinal function of healthy adults. *Int. J. Food Microbiol.* **2006**, *107*, 104–111. [CrossRef] [PubMed]

24. Kim, H.J.; Vazquez Roque, M.I.; Camilleri, M.; Stephens, D.; Burton, D.D.; Baxter, K.; Thomforde, G.; Zinsmeister, A.R. A randomized controlled trial of a probiotic combination VSL# 3 and placebo in irritable bowel syndrome with bloating. *Neurogastroenterol. Motil.* **2005**, *17*, 687–696. [PubMed]

25. Derikx, L.A.; Dieleman, L.A.; Hoentjen, F. Probiotics and prebiotics in ulcerative colitis. *Best Pract. Res. Clin. Gastroenterol.* **2016**, *30*, 55–71. [CrossRef] [PubMed]

26. Bibiloni, R.; Fedorak, R.N.; Tannock, G.W.; Madsen, K.L.; Gionchetti, P.; Campieri, M.; De Simone, C.; Sartor, R.B. VSL#3 probiotic-mixture induces remission in patients with active ulcerative colitis. *Am. J. Gastroenterol.* **2005**, *100*, 1539–1546. [PubMed]

27. Halttunen, T.; Collado, M.C.; El-Nezami, H.; Meriluoto, J.; Salminen, S. Combining strains of lactic acid bacteria may reduce their toxin and heavy metal removal efficiency from aqueous solution. *Lett. Appl. Microbiol.* **2008**, *46*, 160–165. [CrossRef] [PubMed]

28. Nybom, S.M.; Salminen, S.J.; Meriluoto, J.A. Specific strains of probiotic bacteria are efficient in removal of several different cyanobacterial toxins from solution. *Toxicon* **2008**, *52*, 214–220. [CrossRef] [PubMed]

29. El-Nezami, H.; Kankaanpaa, P.; Salminen, S.; Ahokas, J. Ability of dairy strains of lactic acid bacteria to bind a common food carcinogen, aflatoxin B_1. *Food Chem. Toxicol.* **1998**, *36*, 321–326. [CrossRef]

30. Oatley, J.T.; Rarick, M.D.; Ji, G.E.; Linz, J.E. Binding of aflatoxin B_1 to bifidobacteria in vitro. *J. Food Prot.* **2000**, *63*, 1133–1136. [CrossRef] [PubMed]

31. Koepke, R.; Sobel, J.; Arnon, S.S. Global occurrence of infant botulism, 1976–2006. *Pediatrics* **2008**, *122*, E73–E82. [CrossRef] [PubMed]

32. Sakaguchi, G. *Clostridium botulinum* toxins. *Pharmacol. Ther.* **1982**, *19*, 165–194. [CrossRef]

33. Fujinaga, Y. Interaction of botulinum toxin with the epithelial barrier. *J. Biomed. Biotechnol.* **2010**, *210*, 974943. [CrossRef] [PubMed]

34. Couesnon, A.; Molgo, J.; Connan, C.; Popoff, M.R. Preferential entry of botulinum neurotoxin A Hc domain through intestinal crypt cells and targeting to cholinergic neurons of the mouse intestine. *PLoS Pathog.* **2012**, *8*, e1002583. [CrossRef] [PubMed]

35. Connan, C.; Varela-Chavez, C.; Mazuet, C.; Molgo, J.; Haustant, G.M.; Disson, O.; Lecuit, M.; Vandewalle, A.; Popoff, M.R. Translocation and dissemination to target neurons of botulinum neurotoxin type B in the mouse intestinal wall. *Cell. Microbiol.* **2016**, *18*, 282–301. [CrossRef] [PubMed]

36. Matsumura, T.; Sugawara, Y.; Yutani, M.; Amatsu, S.; Yagita, H.; Kohda, T.; Fukuoka, S.; Nakamura, Y.; Fukuda, S.; Hase, K.; et al. Botulinum toxin A complex exploits intestinal M cells to enter the host and exert neurotoxicity. *Nat. Commun* **2015**, *6*. [CrossRef] [PubMed]

37. Ohland, C.L.; Macnaughton, W.K. Probiotic bacteria and intestinal epithelial barrier function. *Am. J. Physiol. Gastrointest. Liver Physiol.* **2010**, *298*, G807–G819. [CrossRef] [PubMed]

38. Hooper, L.V.; Stappenbeck, T.S.; Hong, C.V.; Gordon, J.I. Angiogenins: A new class of microbicidal proteins involved in innate immunity. *Nat. Immunol.* **2003**, *4*, 269–273. [CrossRef] [PubMed]

39. Hooper, L.V.; Wong, M.H.; Thelin, A.; Hansson, L.; Falk, P.G.; Gordon, J.I. Molecular analysis of commensal host-microbial relationships in the intestine. *Science* **2001**, *291*, 881–884. [CrossRef] [PubMed]

40. Anderson, R.C.; Cookson, A.L.; McNabb, W.C.; Kelly, W.J.; Roy, N.C. *Lactobacillus plantarum* DSM 2648 is a potential probiotic that enhances intestinal barrier function. *FEMS Microbiol. Lett.* **2010**, *309*, 184–192. [CrossRef] [PubMed]

41. Hummel, S.; Veltman, K.; Cichon, C.; Sonnenborn, U.; Schmidt, M.A. Differential targeting of the E-cadherin/beta-catenin complex by gram-positive probiotic lactobacilli improves epithelial barrier function. *Appl. Environ. Microbiol.* **2012**, *78*, 1140–1147. [CrossRef] [PubMed]

42. Zyrek, A.A.; Cichon, C.; Helms, S.; Enders, C.; Sonnenborn, U.; Schmidt, M.A. Molecular mechanisms underlying the probiotic effects of *Escherichia coli* Nissle 1917 involve ZO-2 and PKCzeta redistribution resulting in tight junction and epithelial barrier repair. *Cell. Microbiol.* **2007**, *9*, 804–816. [CrossRef] [PubMed]

43. Stetinova, V.; Smetanova, L.; Kvetina, J.; Svoboda, Z.; Zidek, Z.; Tlaskalova-Hogenova, H. Caco-2 cell monolayer integrity and effect of probiotic *Escherichia coli* Nissle 1917 components. *Neuro Endocrinol. Lett.* **2010**, *31*, 51–56. [PubMed]

44. Parassol, N.; Freitas, M.; Thoreux, K.; Dalmasso, G.; Bourdet-Sicard, R.; Rampal, P. *Lactobacillus casei* DN-114 001 inhibits the increase in paracellular permeability of enteropathogenic *Escherichia coli*-infected T84 cells. *Res. Microbiol.* **2005**, *156*, 256–262. [CrossRef] [PubMed]

45. Otte, J.M.; Podolsky, D.K. Functional modulation of enterocytes by gram-positive and gram-negative microorganisms. *Am. J. Physiol. Gastrointest. Liver physiol.* **2004**, *286*, G613–G626. [CrossRef] [PubMed]

46. Carey, C.M.; Kostrzynska, M.; Ojha, S.; Thompson, S. The effect of probiotics and organic acids on Shiga-toxin 2 gene expression in enterohemorrhagic *Escherichia coli* O157:H7. *J. Microbiol. Methods* **2008**, *73*, 125–132. [CrossRef] [PubMed]

47. Rao, R.K.; Samak, G. Protection and restitution of gut barrier by probiotics: Nutritional and clinical implications. *Curr. Nutr. Food Sci.* **2013**, *9*, 99–107. [PubMed]
48. Varankovich, N.V.; Nickerson, M.T.; Korber, D.R. Probiotic-based strategies for therapeutic and prophylactic use against multiple gastrointestinal diseases. *Front. Microbiol.* **2015**, *6*, 685. [CrossRef] [PubMed]
49. Paton, A.W.; Morona, R.; Paton, J.C. Designer probiotics for prevention of enteric infections. *Nat. Rev. Microbiol.* **2006**, *4*, 193–200. [CrossRef] [PubMed]
50. Greenberg, B. *Salmonella* suppression by known populations of bacteria in flies. *J. Bacteriol.* **1969**, *99*, 629–635. [PubMed]
51. Rd, R. Population dynamics of the intestinal tract. In *Colonization Control of Human Bacterial Enteropathogens in Poultry*; Blankenship, L.C., Ed.; Academic Press Inc.: San diego, CA, USA, 1991; pp. 59–75.
52. Chenoll, E.; Casinos, B.; Bataller, E.; Astals, P.; Echevarria, J.; Iglesias, J.R.; Balbarie, P.; Ramon, D.; Genoves, S. Novel probiotic *Bifidobacterium bifidum* CECT 7366 strain active against the pathogenic bacterium *Helicobacter pylori*. *Appl. Environ. Microbiol.* **2011**, *77*, 1335–1343. [CrossRef] [PubMed]
53. Sgouras, D.; Maragkoudakis, P.; Petraki, K.; Martinez-Gonzalez, B.; Eriotou, E.; Michopoulos, S.; Kalantzopoulos, G.; Tsakalidou, E.; Mentis, A. In vitro and in vivo inhibition of *Helicobacter pylori* by *Lactobacillus casei* strain Shirota. *Appl. Environ. Microbiol.* **2004**, *70*, 518–526. [CrossRef] [PubMed]
54. Todoriki, K.; Mukai, T.; Sato, S.; Toba, T. Inhibition of adhesion of food-borne pathogens to Caco-2 cells by lactobacillus strains. *J. Appl. Microbiol.* **2001**, *91*, 154–159. [CrossRef] [PubMed]
55. Chu, H.; Kang, S.; Ha, S.; Cho, K.; Park, S.M.; Han, K.H.; Kang, S.K.; Lee, H.; Han, S.H.; Yun, C.H.; et al. *Lactobacillus acidophilus* expressing recombinant K99 adhesive fimbriae has an inhibitory effect on adhesion of enterotoxigenic *Escherichia coli*. *Microbiol. Immunol.* **2005**, *49*, 941–948. [CrossRef] [PubMed]
56. Tsai, C.C.; Lin, P.P.; Hsieh, Y.M. Three *Lactobacillus* strains from healthy infant stool inhibit enterotoxigenic *Escherichia coli* grown in vitro. *Anaerobe* **2008**, *14*, 61–67. [CrossRef] [PubMed]
57. Munoz, J.A.; Chenoll, E.; Casinos, B.; Bataller, E.; Ramon, D.; Genoves, S.; Montava, R.; Ribes, J.M.; Buesa, J.; Fabrega, J.; et al. Novel probiotic *Bifidobacterium longum* subsp. *infantis* CECT 7210 strain active against rotavirus infections. *Appl. Environ. Microbiol.* **2011**, *77*, 8775–8783. [PubMed]
58. Neeser, J.R.; Granato, D.; Rouvet, M.; Servin, A.; Teneberg, S.; Karlsson, K.A. *Lactobacillus johnsonii* La1 shares carbohydrate-binding specificities with several enteropathogenic bacteria. *Glycobiology* **2000**, *10*, 1193–1199. [CrossRef] [PubMed]
59. Fujiwara, S.; Hashiba, H.; Hirota, T.; Forstner, J.F. Inhibition of the binding of enterotoxigenic *Escherichia coli* Pb176 to human intestinal epithelial cell line HCT-8 by an extracellular protein fraction containing BIF of *Bifidobacterium longum* SBT2928: Suggestive evidence of blocking of the binding receptor gangliotetraosylceramide on the cell surface. *Int. J. Food Microbiol.* **2001**, *67*, 97–106. [PubMed]
60. Mukai, T.; Asasaka, T.; Sato, E.; Mori, K.; Matsumoto, M.; Ohori, H. Inhibition of binding of helicobacter pylori to the glycolipid receptors by probiotic *Lactobacillus reuteri*. *FEMS Immunol. Med. Microbiol.* **2002**, *32*, 105–110. [CrossRef] [PubMed]
61. Hirano, J.; Yoshida, T.; Sugiyama, T.; Koide, N.; Mori, I.; Yokochi, T. The effect of *Lactobacillus rhamnosus* on enterohemorrhagic *Escherichia coli* infection of human intestinal cells *in vitro*. *Microbiol. Immunol.* **2003**, *47*, 405–409. [CrossRef] [PubMed]
62. Coconnier, M.H.; Bernet, M.F.; Chauviere, G.; Servin, A.L. Adhering heat-killed human *Lactobacillus acidophilus*, strain LB, inhibits the process of pathogenicity of diarrhoeagenic bacteria in cultured human intestinal cells. *J. Diarrhoeal Dis. Res.* **1993**, *11*, 235–242. [PubMed]

toxins

MDPI

Review

Clostridium perfringens Sialidases: Potential Contributors to Intestinal Pathogenesis and Therapeutic Targets

Jihong Li [1], Francisco A. Uzal [2] and Bruce A. McClane [1,*]

[1] Department of Microbiology and Molecular Genetics, University of Pittsburgh School of Medicine, Room 420, Bridgeside Point II Building, 450 Technology Drive, Pittsburgh, PA 15219, USA; jihongli@pitt.edu
[2] California Animal Health and Food Safety Laboratory, San Bernardino Branch, School of Veterinary Medicine, University of California-Davis, San Bernardino, CA 92408, USA; fauzal@ucdavis.edu
*   Correspondence: bamcc@pitt.edu; Tel.: +1-412-648-9022

Academic Editor: Holger Barth
Received: 28 October 2016; Accepted: 13 November 2016; Published: 19 November 2016

Abstract: *Clostridium perfringens* is a major cause of histotoxic and intestinal infections of humans and other animals. This Gram-positive anaerobic bacterium can produce up to three sialidases named NanH, NanI, and NanJ. The role of sialidases in histotoxic infections, such as gas gangrene (clostridial myonecrosis), remains equivocal. However, recent in vitro studies suggest that NanI may contribute to intestinal virulence by upregulating production of some toxins associated with intestinal infection, increasing the binding and activity of some of those toxins, and enhancing adherence of *C. perfringens* to intestinal cells. Possible contributions of NanI to intestinal colonization are further supported by observations that the *C. perfringens* strains causing acute food poisoning in humans often lack the *nanI* gene, while other *C. perfringens* strains causing chronic intestinal infections in humans usually carry a *nanI* gene. Certain sialidase inhibitors have been shown to block NanI activity and reduce *C. perfringens* adherence to cultured enterocyte-like cells, opening the possibility that sialidase inhibitors could be useful therapeutics against *C. perfringens* intestinal infections. These initial in vitro observations should be tested for their in vivo significance using animal models of intestinal infections.

Keywords: *Clostridium perfringens*; intestinal infections; gas gangrene; toxins; sialidases; sialidase inhibitors

1. An Introduction to *Clostridium perfringens*

Clostridium perfringens is present throughout the environment, including soil, sewage, feces, foods, and the normal gastrointestinal flora of animals [1,2]. This Gram-positive, anaerobic, spore-forming bacterium is also a feared pathogen of both humans and other animals [2,3]. The most notable *C. perfringens* histotoxic infection is the rapidly-fatal human disease named clostridial myonecrosis (traumatic gas gangrene) [4,5]. This bacterium is also a preeminent cause of common, and sometimes lethal, infections originating in the intestines of humans or livestock [2,6]. Those intestinal infections often involve damage to the small intestine, or to both the small intestine and colon, which results in enteritis or enterocolitis, respectively [1,2]. *C. perfringens* intestinal infections can also progress to enterotoxemia, where a toxin(s) is produced in the intestines and then absorbed to affect extraintestinal organs such as the brain [2,7].

The virulence of this bacterium involves its ability to produce a vast toxin armory [2,3,8]. Currently ~20 different *C. perfringens* toxins have been identified, with more likely awaiting discovery [8–15]. Toxin production repertoires vary greatly among different *C. perfringens* strains, permitting classification of these isolates into five types (A–E), based upon an isolate's production of four typing toxins (alpha, beta, iota, and epsilon toxins) (Table 1) [9,10].

Table 1. *C. perfringens* typing table.

Type	Toxin Production [a]			
	α	β	ε	ι
A	+	−	−	−
B	+	+	+	−
C	+	+	−	−
D	+	−	+	−
E	+	−	−	+

[a] + indicates production of that toxin, while − indicates no production of that toxin.

C. perfringens type designations correlate with disease causation, as shown in Table 2. Two typing toxins, i.e., beta toxin (CPB) and epsilon toxin (ETX), have proven importance in *C. perfringens* intestinal infections of mammalian livestock [3,16,17]. *C. perfringens* produces other toxins that, while not used for typing classification, are nonetheless important for infections originating in the intestines of agriculturally-important animals. The foremost example is necrotic enteritis B (NetB) toxin, which is critical when *C. perfringens* causes avian necrotic enteritis in poultry [12].

Table 2. Diseases associated with the major types/subtypes of *C. perfringens*.

Toxinotypes [a]	Subtype	Most Significant Diseases [b]
A	No CPE [c] or NetB production	Human and animal myonecrosis (gas gangrene)
	NetB-producing	Necrotic enteritis of poultry
	CPE-producing	Human food poisoning and non-foodborne gastrointestinal disease
B		Necro-hemorrhagic enteritis of sheep (lamb dysentery)
C		Human enteritis necroticans (Darmbrand, pigbel); necrotic enteritis of neonatal individuals of several animal species (e.g., cattle, sheep, pigs)
D		Enterotoxemia of sheep and goats
E		Suspected association with gastrointestinal disease of cattle, sheep and rabbits

[a] All types of *C. perfringens* may also produce several other toxins, including, but not limited to, beta2 toxin (CPB2), perfringolysin O (PFO), and toxin *C. perfringens* large cytotoxin (TpeL); [b] Only diseases that have been confirmed to be associated with each type of *C. perfringens* and significant in terms of prevalence are included in this table; [c] CPE is *C. perfringens* enterotoxin.

With respect to human *C. perfringens* infections, type A strains are responsible for causing most histotoxic infections. During gas gangrene, alpha toxin (CPA) plays the major role in virulence. A non-typing toxin named PFO also contributes to this disease [4,5].

To date, only type A and C strains of *C. perfringens* have been conclusively linked to human diseases originating in the intestines [1–3,18]. Type C strains use their CPB to cause enteritis necroticans (EN), which was first described in post-World War II Germany, where it was referred to as darmbrand [18–20]. In the 1960s–1970s, EN, known locally as pigbel, was a major cause of death of children in the Papua New Guinea (PNG) Highlands [20,21]. Pigbel develops in children with reduced trypsin levels due to predisposing conditions, including malnutrition, a diet rich in sweet potato (which contains a trypsin inhibitor), and/or intestinal infections with pathogens producing a trypsin inhibitor [20,21]. Their low intestinal trypsin levels render these children susceptible to infection by type C strains because normal trypsin levels would otherwise easily inactivate CPB when it is produced in the intestines. Consequently, children suffering from pigbel develop CPB-induced necrotic enteritis or enterotoxemia and often die rapidly. The only treatment for pigbel is resection of the bowel; however, this surgical intervention is only effective if performed early after the onset of infection [20,21]. A vaccine introduced in the 1980s dropped the incidence of pigbel dramatically in PNG. Unfortunately, pigbel vaccination has since decreased and this illness may now be reappearing.

Although not used for typing classification, CPE is the toxin responsible for causing the gastrointestinal symptoms of *C. perfringens* type A food poisoning (FP) [1,22]. This FP is currently the 2nd most common bacterial foodborne disease in the USA, where one million cases occur annually and economic losses approach $500 million/year [1,23]. In people with fecal impaction or severe constipation due to side-effects from medications used to treat other pre-existing conditions, *C. perfringens* type A FP can be much more severe and is often fatal [1,24]. Studies with animal models suggest this may be due to absorption of CPE from the intestines, resulting in an enterotoxemia that involves the liver and kidneys [24]. CPE-producing type A strains also cause about 5%–15% of all cases of nonfoodborne human GI diseases, most notably antibiotic-associated diarrhea (AAD) [1,25].

As discussed later, CPE-associated AAD cases are more severe and longer lasting than typical cases of *C. perfringens* type A FP, which usually self-resolve within 24 h [1,25]. While the *cpe* gene can be either chromosomal or plasmid-borne in type A strains, most (~70%) type A FP strains carry a chromosomal *cpe* gene [1]. In contrast, nearly 100% of type A AAD strains carry a plasmid *cpe* gene [1]. There are also many other genetic differences between these two groups of type A *cpe*-positive groups. For example, type A strains carrying a chromosomal *cpe* gene produce a unique small acid soluble protein (SASP4) variant that provides their spores with greater resistance to heat, cold, and chemical treatment and probably facilitates survival of these bacteria in the food environment [1,26,27]. Other differences between type A chromosomal *cpe* strains and type A plasmid *cpe* strains, particularly with regard to sialidase gene carriage, will be discussed later.

In addition to its essential role during intestinal infections by type A *cpe*-positive strains, CPE may also contribute to some cases of human EN caused by type C strains [18]. However, CPB is clearly of critical importance for the pathogenesis of this disease [16].

2. *C. perfringens* Sialidases

Sialic acids are a carbohydrate family containing about 40 different nine-carbon relatives of neuraminic acid [28,29]. Under physiological conditions, sialic acids are negatively charged [29]. N-acetylneuraminic acid (Neu5Ac), whose amino group is acetylated, is the most widespread sialic acid [29,30]. Sialic acids are important components of the serum and mucus and represent the terminal sugar residue of many glycan chains on host cell surfaces, where they are involved in cell- cell recognition. Sialic acids can also stabilize enzymes or cell membrane proteins. Finally, due to their negative charge, sialic acids can mediate binding and transport of positively-charged molecules [28–30].

Sialidases, also referred to as neuraminidases (E.C.3.2.1.18), are key enzymes that hydrolyze the α-linkage of terminal sialic acids on various sialoglycoconjugates to generate free sialic acid [28–31]. Sialidases are made by certain viruses, microorganisms, and vertebrate animals, but not by plants. Included amongst the sialidase-producing viruses and bacteria are several prominent pathogens, e.g., influenza virus, *Vibrio cholerae*, *Streptococcus pneumoniae*, and *C. perfringens*. Sialidases can have a nutritional function for normal flora or pathogenic bacteria [28,29,31]. In addition, they often function directly as virulence factors during bacterial pathogenesis [30,32], as will be discussed later for *C. perfringens*.

C. perfringens produces three different sialidases, which are named NanH, NanI, and NanJ [33]. NanH (43 kDa) lacks a secretion signal peptide and thus has a cytoplasmic location in log-phase cultures [33,34]. In contrast, NanI (77 kDa) and NanJ (129 kDa) are secreted exosialidases ([33], Figure 1). The catalytic modules of all three sialidases show conserved amino acid sequence identity and belong to the family 33 carbohydrate binding module (CBMs, Figure 1) [35]. Compared to NanH, which only consists of a catalytic domain, NanI and NanJ also possess additional accessory carbohydrate-binding modules [35]. It is thought that these carbohydrate-binding domains increase the binding affinity between NanI and NanJ and their polyvalent substrates. NanJ has a complex multimodular structure comprised of a central catalytic module and five accessory modules [35]. The two N-terminal modules show amino acid sequence identity with family 32 and family 40 CBMs. NanI has a simpler structure consisting of a catalytic module and an N-terminal family 40 CBM [35].

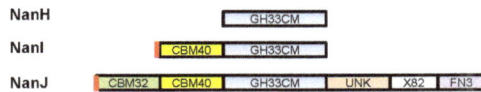

Figure 1. Modular organization of the *C. perfringens* sialidases. CBM32 is a carbohydrate binding module (CBM) belonging to the family 32 CBM; CBM40 is a module having an amino acid sequence identity with a family 40 CBM; GH33CM is a family 33 CBM; UNK is a module having unknown function; X82 is a family 82 "X module" of unknown function; FN3 is a module sharing distant identity with fibronectin type III domains. The red small boxes are secretion signals. Modified with permission from [35]. Copyright 2007 American Chemical Society.

Most *C. perfringens* strains produce all three sialidases (further discussion below). However, as discussed in more detail later, some *C. perfringens* strains produce only one or two of the three sialidases. For strains producing all three sialidases, NanI is usually responsible for ~70% of total exosialidase activity [33,34,36,37].

To characterize the properties of the three *C. perfringens* sialidases, a recent study constructed a series of isogenic mutants, where two of the three sialidase genes present in *C. perfringens* type D strain CN3718 were inactivated [33]. This strategy created mutants that were each expressing, at their native levels, only NanJ, NanI, or NanH in a background free of contamination from the other two sialidases to allow a precise characterization of the enzymatic properties of each sialidase. NanI was found to be more heat-tolerant compared to NanJ or NanH, both of which exhibited greatly reduced sialidase activity at temperatures above 43 °C [33]. In this experimental system, all three sialidases worked best at low pH conditions (pH ~5). The enzyme activity of each sialidase was shown to vary in sensitivity to various metal ions [33]. Furthermore, unlike the sialidases from *Streptomyces* spp., *C. perfringens* sialidases were found to be sensitive to p-chloromercuribenzoate, which reacts with thiol groups in proteins [33]. Finally, the three *C. perfringens* sialidases showed different substrate preferences. NanI exhibited preferential activity in the order of α-2,3 > α-2,6 > α-2,8 sialic acid linkages and was responsible for most of the activity in CN3718 supernatants that was directed against those sialic acid linkages. NanJ showed a preference for α-2,6 > α-2,8 > α-2,3 sialic acid linkages. Finally, NanH activity was strongest for α-2,8 > α-2,3 > α-2,6 sialic acid linkages [33]. This diversity in linkage preferences suggests that, when present together, the three *C. perfringens* sialidases work in combination to generate free sialic acid, even from complex substrates [33].

3. *C. perfringens* Sialidases: Genetics and Regulation of Expression

All three *C. perfringens* sialidases are encoded by chromosomal genes, although those genes are located in different regions of the chromosome [38,39]. The sialidase-encoding ORFs in many *C. perfringens* strains have now been sequenced. Those sequencing analyses indicated that, amongst different *C. perfringens* strains, the NanJ sequence shares 96% to 100% identity, the NanI sequence has 98% to 100% identity, and the NanH sequence shares 93% to 100% identity [34,38,39].

As true for their toxin production, *C. perfringens* strains vary in their patterns of sialidase production. Most strains produce all three sialidases, with NanI usually being responsible for most of the sialidase activity in culture supernatants of those *C. perfringens* strains [34,36]. However, NanI production is not essential for *C. perfringens* growth since some strains of this bacterium naturally lack the *nanI* gene [36]. Interestingly, the *nanI* gene is consistently absent from the type A FP strains carrying a chromosomal *cpe* gene, as well as the genetically related type C darmbrand strains [36]. In contrast, the *nanI* gene is carried by most plasmid *cpe*-carrying type A AAD strains, *cpe*-negative type A normal human intestinal flora strains, and type C pig-bel strains [36]. Since NanI is usually the major sialidase of *C. perfringens*, it is not surprising that exosialidase activity is typically significantly lower for naturally *nanI*-negative strains compared to strains carrying a *nanI* gene [36]. The potential pathogenic importance of these differences will be discussed later.

Most type A FP strains with a chromosomal *cpe* gene lack the *nanJ* gene, as well as the *nanI* gene. However, those typical FP strains do carry the *nanH* gene. It should be noted that occasional strains of *C. perfringens* besides the type A chromosomal *cpe* FP strains or type C darmbrand strains also lack a sialidase gene. For example, Strain 13 is a *cpe*-negative type A strain that can cause gas gangrene yet it lacks the *nanH* gene [37,40].

The regulation of sialidase production by *C. perfringens* is complicated (Figure 2). A number of regulators influencing expression of one or more sialidase genes have been identified. For example, the VirS/VirR two-component signal transduction system was shown to upregulate *nanI* and *nanJ* expression [41]. Rather than a direct regulatory effect involving VirS binding to sialidase gene promoters, this VirS/R two-component system positively controls expression of the *vrr* gene, which encodes *virR*-regulated RNA (VR-RNA). This regulatory RNA then modulates sialidase gene expression [42]. ReeS is another sensor kinase whose presence also increases *nanI* and *nanJ* sialidase gene expression by a putative response regulator, referred to as ReeR [43]. In contrast, a transcriptional regulator named RevR has different regulatory effects on *nanJ* and *nanI* expression, possibly via a proposed sensor kinase named RevS. RevR increases *nanJ* expression, but negatively regulates *nanI* expression [44]. Finally, using a *codY* null mutant of *C. perfringens* type D strain CN3718, our group showed that CodY represses NanJ and NanH production, although it does not affect NanI production [45].

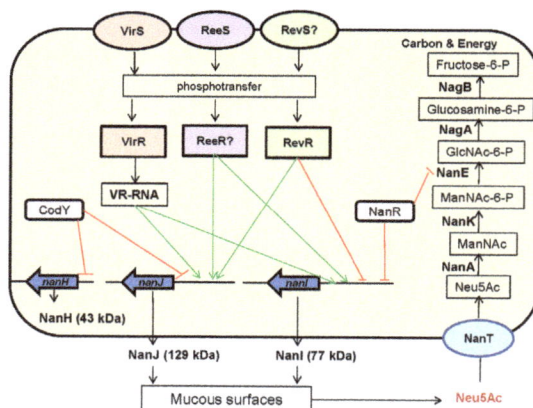

Figure 2. A proposed model for regulation of expression of sialidase genes and the pathway for sialic acid metabolism in *C. perfringens*. Exosialidases generate free sialic acid from mucus or host cell surfaces [33,40]. The free sialic acid is then transported into *C. perfringens* where it is metabolized to fructose-6-P [40]. There is evidence that VirS/VirR, RevR, ReeS, NanR, and CodY systems directly or indirectly affect sialidase production, although inter-relationships between these regulators are unclear [40,41,43–46]. The VirS/VirR two component system acts as a positive regulator of the *vrr* gene, which encodes VR-RNA. VR-RNA is then a positive regulator of *nanI* and *nanJ* expression [41,42]. The ReeS sensor kinase positively regulates *nanI* and *nanJ* gene expression, presumably by a putative transcriptional regulator named ReeR [43]. RevS positively regulates *nanJ* expression but negatively regulates *nanI* expression [44]. CodY represses *nanH* and *nanJ* expression [45]. Based on sequence homology comparisons with similar regulators in other bacteria, NanR may repress *nanI* expression and the sialic acid metabolism pathway, but there is no direct experimental evidence yet to support this hypothesis. Green lines indicate positive regulation while red lines indicate negative regulation.

No proven regulator has yet been identified that directly modulates sialidase production by binding to the promoters of *C. perfringens* sialidase genes. However, gel mobility shift assays have demonstrated high affinity binding of a purified protein named NanR to DNA from the promoter region of the *nanI* gene [40]. Since NanR has homology with the ribose-5-phosphate isomerase B regulator (RpiR) family of transcriptional repressors known to control sialidase production in bacteria such as *E. coli, Vibrio vulnificus*, and *Staphylococcus aureus*, this result suggests that NanR may be involved in *nanI* expression regulation [31,40,47]. Consistent with its potential involvement in sialic acid generation and usage, NanR lies within a six-gene operon encoding the complete pathway for transport and metabolism of sialic acid by *C. perfringens* [40]. As discussed in more detail below, this operon also encodes NanE (epimerase) and NanA (sialic acid lyase) enzymes.

For the *nanI* and *nanJ* genes, primer extension analyses identified three or two putative transcription start sites, respectively [40]. These promoters are located within ~500 bp of the start codons of *nanI* and *nanJ*. This multiplicity of promoters may provide one explanation for why so many different regulators control sialidase expression, although (as mentioned) detailed understanding of regulator binding to these promoter sequences is currently lacking [40].

Only a limited number of bacterial species, mainly those having a close association with the sialic acid-rich environment of the host, can utilize Neu5Ac [48]. *C. perfringens* was actually the first bacterium demonstrated to be capable of utilizing a sialic acid (Neu5Ac) as a carbon source [31]. This result was later confirmed by another study showing that Neu5Ac can be used by *C. perfringens* when growing in a semi-defined medium [31]. It was also recently found that NanI and NanJ, but not NanH, can cause the release of sialic acid from Caco-2 cells [33]. This effect may contribute to pathogenesis, since these sialidases could help *C. perfringens* obtain nutrients in vivo by releasing sialic acid from glycolipids or glycoproteins on the host cell surface or in mucus. Interestingly, contact with Caco-2 cells was shown to upregulate the expression of NanI [34], which could potentiate NanI contributions during intestinal infections, as discussed later.

NanI, but not NanJ, production is induced by the addition of Neu5Ac to a medium [40]. As introduced earlier, NanR is part of an operon that encodes a complete pathway for the transport and metabolism of sialic acid. After sialic acid is generated by sialidases, it is then transported and metabolized by products of this operon, also referred to as the Nan cluster. This process involves an initial conversion of sialic acid to N-acetyl glucosamine, followed by metabolism of that carbohydrate to fructose-6-P, which *C. perfringens* can use as carbon sources and for energy production (Figure 2) [40,46].

Many other pathogenic and commensal bacteria found in the intestines carry a Nan cluster for sialic acid utilization [48]. Examples of such bacteria include *Vibrio cholerae* and *Salmonella enterica* [48,49]. Many of these bacteria capable of utilizing sialic acid also colonize the human or animal intestinal tract. This correlation is probably not a coincidence since sialic acid is a component of mucin, the major protein in mucus, which is abundant in the human and animal intestines [48]. However, to generate free sialic acid from mucus for uptake and metabolism inside the bacterial cell, extracellular sialidases must be present in the intestinal environment. Interestingly, some pathogenic bacteria, e.g., *C. difficile*, have a Nan cluster but do not produce their own sialidase. Instead, it is believed that *C. difficile* uses free sialic acid generated in the intestines by sialidases produced by other bacteria, such as *Bacteroides thetaiotaomicron* [50,51].

4. Possible Contributions of Sialidases to *C. perfringens* Diseases

Some pathogens use sialic acids to coat their cell surface, the flagellum, the capsule polysaccharide, or the lipopolysaccharide. This masks these bacteria so they can avoid the host immune system defense [28]. Whether *C. perfringens* coats its surface with host-derived sialic acid has not been studied, to our knowledge.

Sialidases can also promote in vivo growth and colonization of bacterial pathogens [30]. For extraintestinal infections, an example is the major human respiratory tract pathogen *Streptococcus pneumoniae*, which encodes up to three sialidases, named NanA, NanB, and NanC. NanA is the predominant sialidase that removes the sialic acid Neu5Ac from a variety of glycoconjugates. NanB is an intramolecular trans-sialidase producing 2,7-anhydro-Neu5Ac selectively from α2,3-sialosides, while NanC produces 2-deoxy-2,3-didehydro-*N*-acetylneuraminic acid (Neu5Ac2en), which can be hydrated to Neu5Ac. The three pneumococcal sialidases share a common catalytic mechanism up to the final product formation step, and all three sialidases are implicated in pathogenesis, including colonization, and are potential drug targets [52]. *S. pneumoniae* then takes up and metabolizes sialic acid using a similar pathway as present in *C. perfringens* [40].

The contribution of sialidases to growth and colonization during extraintestinal infections caused by *C. perfringens* is unclear. One study demonstrated that a *nanJ* and *nanI* double null mutant of *C. perfringens* strain 13 remains fully virulent in the mouse myonecrosis model [37]. This result could suggest that Neu5Ac metabolism is not essential for growth or colonization by *C. perfringens* in muscle. However, as noted in that study [37], this result does not necessarily preclude subtle contributions of sialidases to colonization or growth since the mouse myonecrosis model requires a massive inoculum that may mask such contributions.

Increasing evidence indicates that sialidases and sialic acid often play significant roles in growth and colonization of the intestines by bacterial pathogens [34,49,52,53]. For example, sialic acid appears to be an important source of carbon and energy for survival, growth and adherence of *E. coli* in the gastrointestinal system [54]. Similarly, catabolism of sialic acid by *V. cholerae* plays a significant role in both in vitro and in vivo colonization and growth [49]. In addition, *V. cholerae* sialidase enhances the binding and uptake of cholera toxin [55]. In *V. vulnificus*, a Nan utilization system is also important for bacterial colonization and growth in the intestines [56]. The sialidase NanS play a role in *Clostridium sordellii* adhesion and enhances Non-TcsL mediated cytotoxicity [53]. As mentioned above, *C. difficile* uses sialic acid generated by sialidases of other bacteria to grow and persist in the intestines [50,51].

As mentioned earlier, NanI may contribute to in vivo growth. Contact of *C. perfringens* with Caco-2 cells increases the production of this enzyme [34] and NanI can induce the release of free sialic acids from enterocyte-like Caco-2 cells [33]. Since *C. perfringens* possesses a sialic acid utilization system, similar NanI-generated sialic acid in the intestines should be useful for in vivo growth of this bacterium.

Earlier studies demonstrated that NanI production also impacts the production of several *C. perfringens* toxins [34,37,57]. Specifically, inactivating the *nanI* gene in *C. perfringens* type A strain 13 caused a slight increase in supernatant activities of alpha toxin and perfringolysin O (PFO), which are contributors to gas gangrene [37], but their role (if any) in most intestinal infections, particularly human infections, is unsettled [3]. However, it was later demonstrated that inactivating the *nanI* gene in type D strain CN3718 reduced ETX toxin production in vitro and that either genetic or physical complementation could recover this ETX production [57]. NanI effects on ETX production involved reductions in both *codY* and *ccpA* transcript levels, suggesting a model whereby NanI generates sialic acid release in the intestines and that free sialic acid then alerts a type D strain of its presence in the intestines, making it worthwhile for this bacterium to upregulate ETX production to induce disease.

There have been only limited studies to date on the effects of *C. perfringens* sialidase on toxin action, particularly on those toxins important for intestinal infections. An exception is ETX, where conclusions from early studies had been contradictory. One early study [58] reported that sialidases enhance ETX cytotoxicity towards Madin-Derby Canine Kidney (MDCK) epithelial cells, while a second study [59] reported that pretreating synaptosomal membranes with sialidases lowers their subsequent ETX binding levels. Recently a study used purified NanI, as well as *nanI, nanJ,* or *nanH* single null mutants, a *nanI/nanJ* double null mutant, and a triple sialidase null mutant, to study possible sialidase enhancement of ETX action [34]. Results obtained [34] conclusively showed that NanI sialidase increases the ETX sensitivity of MDCK cells and that this effect involves an increase in ETX binding levels. The mechanism of this enhancement is not yet clear but it could involve either NanI increasing the exposure of ETX receptors on the host cell surface or NanI modifying the host cell surface charge to increase binding of this toxin. Whether a similar effect extends to other *C. perfringens* toxins active during intestinal infections is not yet clear.

Only limited studies have addressed the mechanism of *C. perfringens* adherence to host cells and tissues [60–62]. For example, the adhesins used by *C. perfringens* in vivo remain unclear. However, some strains produce a collagen adhesion protein (CNA), and/or fibronectin binding proteins (FbpA, FbpB) that have been implicated in adhesion and colonization [60–63].

Sialidases are known to contribute to the ability of some bacterial pathogens to adhere to host cells, tissues, or mucosal surfaces in the airways or intestines [30]. A recent in vitro study suggests sialidase contributions to *C. perfringens* intestinal adherence [34]. That study [34] demonstrated that type D strain CN3718 attaches to cultured Caco-2 enterocyte-like cells. This adhesion was specific since this intestinal disease strain exhibited significantly less adherence to fibroblasts and kidney cells [34]. Interestingly, the attachment of CN3718 to Caco-2 cells is greatly facilitated by NanI production (Figure 3). Wild-type CN3718 exhibited much more adherence to Caco-2 cells compared to an isogenic triple mutant strain that does not produce any sialidase. Complementation studies showed that, of the three sialidases made by CN3718, restoring NanI production to the triple sialidase mutant yielded the greatest enhancement of adherence. That result indicated that NanI is of prime importance for sialidase enhancement of CN3718 attachment to Caco-2 cells [34].

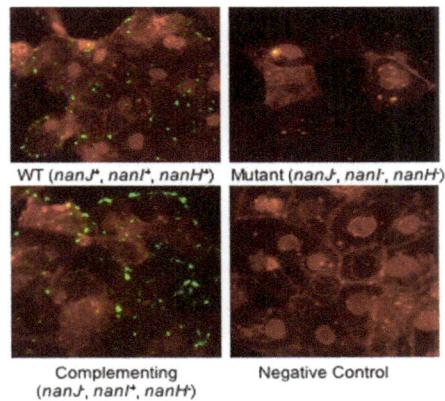

WT (*nanJ+, nanI+, nanH+*)   Mutant (*nanJ-, nanI-, nanH-*)

Complementing
(*nanJ-, nanI+, nanH-*)   Negative Control

Figure 3. Adhesion of *C. perfringens* CN3718 to Caco-2 cells [34]. Caco-2 cells were incubated for 2 h at 37 °C under anaerobic conditions. *C. perfringens* CN3718 produces all three sialidases. This wild-type (WT) strain, and a complementing strain that produces only NanI, attach very well to Caco-2 cells as detected by immunofluorescence microscopy (600×). In comparison, only a few cells of an isogenic mutant with all three sialidase genes disrupted were able to attach to Caco-2 cells. Furthermore, when the triple mutant was complemented to produce only NanJ or NanH, those bacteria remained poorly adherent (not shown). Green: *C. perfringens*; Red: Caco-2 cells. Reproduced with permission from [34]. Creative Commons License 2011, Copyright J. Li and B.A. McClane.

However, NanI itself does not appear to be a significant adhesin since the CN3718 *codY* null mutant shows 50% less adherence to Caco-2 cells compared to wild-type CN3718, despite this mutant producing the same levels of NanI as the wild-type strain [45]. This result suggests a two-step adherence mechanism. First, secreted NanI modifies the Caco-2 enterocyte-cell surface, which then allows the unknown adhesin present on the surface of *C. perfringens* to more easily bind to the still unidentified receptor(s) on the enterocyte-cell surface.

Sialidases may enhance *C. perfringens* adherence to host cells via both nonspecific and specific mechanisms. Sialic acids, typically present at the distal ends of carbohydrate chains, are negatively charged carbohydrates [28,30]. In addition, terminal sialic acids promote endothelial barrier integrity, so treatment of epithelial monolayers with *C. perfringens* sialidases may lead to barrier disruption and may increase access for *C. perfringens* so they can adhere [64]. Therefore, nonspecific effects of secreted NanI on both charge and epithelial barrier integrity could help to increase toxin binding and *C. perfringens* colonization. A *C. perfringens* sialidase (not specified) has, in fact, been shown to affect barrier resistance in some, but not all, host cells [64]. However, for some cell lines, nonspecific effects do not appear to completely explain NanI enhancement of *C. perfringens* adherence. As mentioned, *C. perfringens* CN3718 exhibits much greater adhesion for certain mammalian cells, such as Caco-2 and HT-29 intestinal cell lines, than for cell lines of nonintestinal origin [34]. Similarly, the toxins involved in *C. perfringens* intestinal infections only bind to and affect certain cells. Those observations suggest that NanI can also have specific effects in promoting *C. perfringens* adherence and toxin binding. This could involve specific effects of NanI sialidase on modifying host cell surface adhesins/toxin receptors and/or trimming back nearby molecules to better unmask the adhesin or toxin receptor on host cell surfaces.

When *C. perfringens* causes diseases originating in the intestines, its secreted proteins come into contact with host proteases, such as trypsin, that are present in the intestinal lumen. While some *C. perfringens* proteins, e.g., CPB, are highly sensitive to intestinal proteases [16], other proteins secreted by this bacterium are proteolytically-activated in the intestines. A prime example of a proteolytically-activated *C. perfringens* protein is ETX, which is initially produced as an inactive prototoxin and then activated by intestinal trypsin, chymotrypsin, and carboxypeptidases that remove N- and C-terminus residues from the prototoxin [65]. Similarly, trypsin or chymotrypsin treatment of CPE also increases cytotoxicity, in this case by removing N-terminal sequences from the native CPE protein to facilitate toxin oligomerization during pore formation [66]. Interestingly, NanI (but not NanJ or NanH) can also be proteolytically-activated by trypsin [33,34]. The increased enzymatic activity of trypsin-activated NanI was shown to be substrate-specific [33]. Collectively, these observations suggest that trypsin activation of NanI may contribute to *C. perfringens* intestinal diseases.

While NanI can enhance *C. perfringens* adherence and toxin binding, typical type A FP strains, i.e., those with a chromosomal *cpe* gene, do not carry the *nanI* gene. Furthermore, those typical FP strains possess very low exosialidase activities [36]. The limited production of sialidases by type A FP strains obviously does not hinder their ability to cause food-borne intestinal diseases. Unlike the typical FP strains, nearly all type A AAD strains do carry a *nanI* gene. When a *nanI* null mutant of a type A AAD strain was characterized, this *nanI* null mutant was shown to enter the sporulation cycle earlier and to produce more CPE than its wild-type parent. This finding supports the dispensability of NanI for the typical FP strains, which cause an acute disease. During the acute FP, NanI would not be needed for nutritional purposes or to enhance bacteria adherence or colonization since the typical FP strains are ingested in very large numbers in contaminated food, then quickly sporulate in the intestines to produce CPE, induce diarrhea via that CPE, and exit from the intestines via the flushing action of diarrhea (Figure 4) [36]. In contrast, type A AAD or sporadic diarrhea (SD) strains with a plasmid *cpe* gene cause chronic gastrointestinal disease lasting up to several weeks, so in vivo growth and colonization is important for their persistence. Those Type A AAD or SD strains may use NanI for their growth and colonization, as necessary to achieve the persistence required to cause a chronic diarrhea (Figure 4) [36].

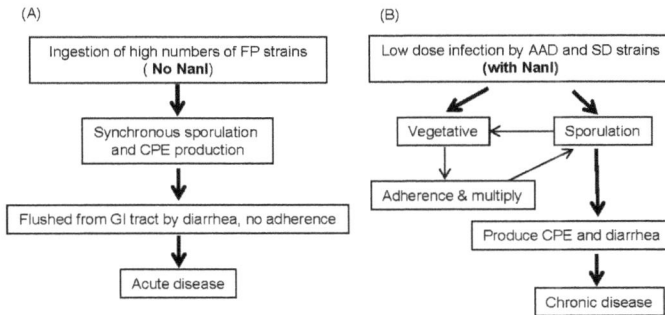

Figure 4. Possible models for acute *C. perfringens* type A FP vs. chronic gastrointestinal diseases like AAD. (**A**) *C. perfringens* type A FP does not require NanI production because strains are ingested in large amounts, sporulate in vivo to produce CPE and are then quickly removed from the intestines by diarrhea. (**B**) *C. perfringens* type A AAD does involve NanI production, which may promote adherence and colonization, as required for chronic diarrhea. Modified with permission from [36]. Copyright 2014 American Society for Microbiology.

5. Sialidase Inhibitors: Potential Therapeutic Agents?

As described above, sialidases may offer several contributions to *C. perfringens* pathogenesis, particularly during intestinal infections. *C. perfringens* diseases are challenging to treat with antibiotics because already-synthesized toxins will continue to work even after the administration of antibiotics. Similarly, it is often difficult to treat or prevent *C. perfringens* infections with vaccines or neutralizing antibodies because some individual strains produce multiple (up to five) different toxins [2,3]. Those factors suggest that sialidase inhibitors represent interesting potential candidates for drug development.

Sialidase inhibitors have proven to be efficacious for the treatment of some infectious disease [67]. The most notable clinical application of sialidase inhibitors are several sialidase inhibitors (Zanamivir, Oseltamivir, Peramivir (Rapivab)) that are currently used for treating infections involving influenza virus [67]. However, there is also precedent for sialidase inhibitors interfering with bacterial growth and adhesion. Oseltamivir reduced the growth and adherence of *Tannerella forsythia*, which causes periodontitis, by two- to three-fold [68].

Two classic sialidase inhibitors, i.e., Siastatin B (SB) and N-acetyl-2,3-dehydro-2-deoxyneuraminic acid (NADNA), have been tested against *C. perfringens* [33,36,57,69,70]. SB, a broad-spectrum sialidase inhibitor isolated from a *Streptomyces* spp. culture, is an unusual 6-acetamido-3-piperidinecarboxy late [70]. NADNA is an analogue of 2-deoxy-2,3-didehydro-N-acetylneuraminic acid that is modified at the C-4 position. NADNA can inhibit the sialidase activity from influenza viruses A and B, parainfluenza 2 virus, *Vibrio cholerae*, *Arthrobacter ureafaciens*, *C. perfringens*, and sheep liver [69].

A recent study measured the effects of SB and NADNA on sialidase activities of overnight culture supernatants of *C. perfringens* CN3718 mutants ENanJ, ENanI, or ENanH, which contain natural levels of only NanJ, NanI, or NanH, respectively [33]. NanI was the most sensitive sialidase to both NADNA and SB, while NanH was the least sensitive sialidase to these inhibitors; IC_{50}s for these inhibitors are shown in Table 3 [33].

Table 3. Effect of sialidase inhibition on sialidase activity (IC$_{50}$).

Sample	NADNA (IC$_{50}$)	Siastatin B (IC$_{50}$)
CN3718	18.9 μM	42.2 μM
ENanJ	12.4 μM	15.1 μM
ENanI	13.4 μM	27.5 μM
ENanH	44.6 μM	50.9 μM

It was also shown that the adherence of *C. perfringens* AAD strain F4969 to Caco-2 cells can be reduced by either SB or NADNA [36]. In addition, both NADNA and SB efficiently inhibited sialidase activity in bacterial cell-free supernatants collected from *C. perfringens* type D strain CN3718. Consistent with those results, CN3718 cultures grown in the presence of the SB inhibitor also exhibited substantially reduced culture sialidase activity and ETX production. However, the NADNA sialidase inhibitor did not inhibit sialidase activity in CN3718 cultures, even at very high doses of the inhibitor. In the presence of NADNA, CN3718 cultures still possessed strong sialidase activity and also made the same amounts of ETX as wild-type CN3718 grown in the absence of any sialidase inhibitor [57]. Collectively, these results suggest that at least some sialidase inhibitors may be potentially useful therapeutics against *C. perfringens* infections.

6. Concluding Remarks and Future Directions

Most studies of *C. perfringens* virulence have, correctly, focused on toxin contributions. Only a few in vivo studies have thus far examined the role of exoenzymes in *C. perfringens* pathogenesis and their results have been equivocal. However, those studies examined the possible role of sialidases in gas gangrene. Recent in vitro studies suggest that sialidases, particularly NanI, may be important contributors to *C. perfringens* intestinal infections. Those contributions could include upregulated production of some toxins relevant to intestinal infections, enhanced binding and activity of some of those toxins, increased adherence of *C. perfringens* to host cells, and possibly generation of substrates for growth and metabolism [33,34,36,57]. There is a clear need to move these promising findings into appropriate animal models to firmly evaluate sialidase contributions to *C. perfringens* virulence.

In vitro studies also suggest that some sialidase inhibitors might be useful therapeutics for treating *C. perfringens* infections originating in the intestines [57]. This possibility also needs to be evaluated in animals. Identification of more potent inhibitors of *C. perfringens* sialidases may be helpful for those studies.

Acknowledgments: Preparation of this review was supported, in part, by a grant from the National Institute of Allergy and Infectious Disease (R21 AI125796-1).

Conflicts of Interest: The authors declare no conflict of interest.

References

1. McClane, B.A.; Robertson, S.L.; Li, J. *Clostridium perfringens*. In *Food Microbiology: Fundamentals and Frontiers*, 4th ed.; Doyle, M.P., Buchanan, R.L., Eds.; ASM Press: Washington, DC, USA, 2013; pp. 465–489.
2. McClane, B.A.; Uzal, F.A.; Miyakawa, M.F.; Lyerly, D.; Wilkins, T.D. The enterotoxic clostridia. In *The Prokaryotes*, 3rd ed.; Dworkin, M., Falkow, S., Rosenburg, E., Schleifer, H., Stackebrandt, E., Eds.; Springer: New York, NY, USA, 2006; pp. 688–752.
3. Uzal, F.A.; Freedman, J.C.; Shrestha, A.; Theoret, J.R.; Garcia, J.; Awad, M.M.; Adams, V.; Moore, R.J.; Rood, J.I.; McClane, B.A. Towards an understanding of the role of *Clostridium perfringens* toxins in human and animal disease. *Future Microbiol.* **2014**, *9*, 361–377. [CrossRef] [PubMed]
4. Rood, J.I. *Clostridium perfringens* and histotoxic disease. In *The prokaryotes: A Handbook on the Biology of Bacteria*, 3rd ed.; Dworkin, M., Falkow, S., Rosenberg, E., Schleifer, K.-H., Stackebrandt, E., Eds.; Springer: New York, NY, USA, 2007; Volume 4, pp. 753–770.

5. Titball, R.W.; Rood, J.I. *Clostridium pefringens*: Wound indections. In *Molecular Medical Microbiology*; Sussman, M., Ed.; Academic Press: London, UK, 2002; pp. 1875–1903.

6. Uzal, F.A.; Vidal, J.E.; McClane, B.A.; Gurjar, A.A. *Clostridium perfringens* toxins involved in mammalian veterinary diseases. *Open toxinol. J.* **2010**, *2*, 24–42. [CrossRef]

7. Songer, J.G. Clostridial enteric diseases of domestic animals. *Clin. Microbiol. Rev.* **1996**, *9*, 216–234. [PubMed]

8. Hatheway, C. Toxigenic clostridia. *Clin. Microb. Rev.* **1990**, *3*, 66–76. [CrossRef]

9. McDonel, J.L. Toxins of *Clostridium perfringens* types A, B, C, D, and E. In *Pharmacology of Bacterial Toxins*; Dorner, F., Drews, H., Eds.; Pergamon Press: Oxford, UK, 1986; pp. 477–517.

10. Petit, L.; Gilbert, M.; Popoff, M. *Clostridium perfringens*: Toxinotype and genotype. *Trends Microbiol.* **1999**, *7*, 104–110. [CrossRef]

11. Animoto, K.; Noro, T.; Oishi, E.; Shimizu, M. A novel toxin homologous to large clostridial cytotoxins found in culture supernatant of *Clostridium perfringens* type C. *Microbiology* **2007**, *153*, 1198–1206. [CrossRef] [PubMed]

12. Keyburn, A.L.; Boyce, J.D.; Vaz, P.; Bannam, T.L.; Ford, M.E.; Parker, D.; Di Rubbo, A.; Rood, J.I.; Moore, R.J. NetB, a new toxin that is associated with avian necrotic enteritis caused by *Clostridium perfringens*. *PLoS Pathog.* **2008**, *4*, e26. [CrossRef] [PubMed]

13. Mehdizadeh Gohari, I.; Parreira, V.R.; Timoney, J.F.; Fallon, L.; Slovis, N.; Prescott, J.F. NetF-positive *Clostridium perfringens* in neonatal foal necrotising enteritis in kentucky. *Vet. Rec.* **2016**, *178*, 216. [CrossRef] [PubMed]

14. Yonogi, S.; Matsuda, S.; Kawai, T.; Yoda, T.; Harada, T.; Kumeda, Y.; Gotoh, K.; Hiyoshi, H.; Nakamura, S.; Kodama, T.; et al. BEC, a novel enterotoxin of *Clostridium perfringens* found in human clinical isolates from acute gastroenteritis outbreaks. *Infect Immun.* **2014**, *82*, 2390–2399. [CrossRef] [PubMed]

15. Irikura, D.; Monma, C.; Suzuki, Y.; Nakama, A.; Kai, A.; Fukui-Miyazaki, A.; Horiguchi, Y.; Yoshinari, T.; Sugita-Konishi, Y.; Kamata, Y. Identification and characterization of a new enterotoxin produced by *Clostridium perfringens* isolated from food poisoning outbreaks. *PLoS ONE* **2015**, *10*, e0138183. [CrossRef] [PubMed]

16. Sayeed, S.; Uzal, F.A.; Fisher, D.J.; Saputo, J.; Vidal, J.E.; Chen, Y.; Gupta, P.; Rood, J.I.; McClane, B.A. Beta toxin is essential for the intestinal virulence of *Clostridium perfringens* type C disease isolate CN3685 in a rabbit ileal loop model. *Mol. Microbiol.* **2008**, *67*, 15–30. [CrossRef] [PubMed]

17. Garcia, J.P.; Adams, V.; Beingesser, J.; Hughes, M.L.; Poon, R.; Lyras, D.; Hill, A.; McClane, B.A.; Rood, J.I.; Uzal, F.A. Epsilon toxin is essential for the virulence of *Clostridium perfringens* type D infection in sheep, goats and mice. *Infect Immun.* **2013**, *81*, 2405–2414. [CrossRef] [PubMed]

18. Ma, M.; Gurjar, A.; Theoret, J.R.; Garcia, J.P.; Beingesser, J.; Freedman, J.C.; Fisher, D.J.; McClane, B.A.; Uzal, F.A. Synergistic effects of *Clostridium perfringens* enterotoxin and beta toxin in rabbit small intestinal loops. *Infect Immun.* **2014**, *82*, 2958–2970. [CrossRef] [PubMed]

19. Ma, M.; Li, J.; McClane, B.A. Genotypic and phenotypic characterization of *Clostridium perfringens* isolates from darmbrand cases in post-World War II Germany. *Infect Immun.* **2012**, *80*, 4354–4363. [CrossRef] [PubMed]

20. Lawrence, G.W. The pathogenesis of enteritis necroticans. In *The Clostridia: Molecular Genetics and Pathogenesis*; Rood, J.I., McClane, B.A., Songer, J.G., Titball, R.W., Eds.; Academic Press: London, UK, 1997; pp. 198–207.

21. Johnson, S.; Gerding, D.N. Enterotoxemic infections. In *The Clostridia: Molecular Biology and Pathogenesis*; Rood, J.I., McClane, B.A., Songer, J.G., Titball, R.W., Eds.; Academic Press: London, UK, 1997; pp. 117–140.

22. Freedman, J.C.; Shrestha, A.; McClane, B.A. *Clostridium perfringens* enterotoxin: Action, genetics, and translational applications. *Toxins* **2016**, *8*. [CrossRef] [PubMed]

23. Centers for Disease Control and Prevention (CDC). Posting date. CDC estimates of foodborne illness in the United States: *Clostridium perfringens*. Available online: http://www.cdc.gov/foodsafety/diseases/clostridium-perfringens.html (accessed on 28 July 2016).

24. Caserta, J.A.; Robertson, S.L.; Saputo, J.; Shrestha, A.; McClane, B.A.; Uzal, F.A. Development and application of a mouse intestinal loop model to study the in vivo action of *Clostridium perfringens* enterotoxin. *Infect. Immun.* **2011**, *79*, 3020–3027. [CrossRef] [PubMed]

25. Carman, R.J. *Clostridium perfringens* in spontaneous and antibiotic-associated diarrhoea of man and other animals. *Rev. Med. Microbiol.* **1997**, *8* (Suppl. S1), S43–S45. [CrossRef]

26. Li, J.; Paredes-Sabja, D.; Sarker, M.; Mcclane, B. *Clostridium perfringens* sporulation and sporulation-associated toxin production. *Microbiol. Spectr.* **2016**, *4*. [CrossRef]

27. Li, J.; McClane, B.A. A novel small acid soluble protein variant is important for spore resistance of most *Clostridium perfringens* food poisoning isolates. *PLoS Pathog.* **2008**, *4*, e1000056. [CrossRef] [PubMed]

28. Severi, E.; Hood, D.W.; Thomas, G.H. Sialic acid utilization by bacterial pathogens. *Microbiology* **2007**, *153*, 2817–2822. [CrossRef] [PubMed]

29. Traving, C.; Schauer, R. Structure, function and metabolism of sialic acids. *Cell. Mol. Life Sci.* **1998**, *54*, 1330–1349. [CrossRef] [PubMed]

30. Lewis, A.L.; Lewis, W.G. Host sialoglycans and bacterial sialidases: A mucosal perspective. *Cell. Microbiol.* **2012**, *14*, 1174–1182. [CrossRef] [PubMed]

31. Vimr, E.R.; Kalivada, K.A.; Deszo, E.L.; Steenbergen, S.M. Diversity of microbial sialic acid metabolism. *Microbiol. Molec Biol. Rev.* **2004**, *68*, 132–153. [CrossRef]

32. Rohmer, L.; Hocquet, D.; Miller, S.I. Are pathogenic bacteria just looking for food? Metabolism and microbial pathogenesis. *Trends Microb.* **2011**, *19*, 341–348. [CrossRef] [PubMed]

33. Li, J.; McClane, B.A. The sialidases of *Clostridium perfringens* type D strain CN3718 differ in their properties and sensitivities to inhibitors. *Appl. Environ. Microbiol.* **2014**, *80*, 1701–1709. [CrossRef] [PubMed]

34. Li, J.; Sayeed, S.; Robertson, S.; Chen, J.; McClane, B.A. Sialidases affect the host cell adherence and epsilon toxin-induced cytotoxicity of *Clostridium perfringens* type D strain CN3718. *PLoS Pathog.* **2011**, *7*, e1002429. [CrossRef] [PubMed]

35. Boraston, A.B.; Ficko-Blean, E.; Healey, M. Carbohydrate recognition by a large sialidase toxin from *Clostridium perfringens*. *Biochemistry* **2007**, *46*, 11352–11360. [CrossRef] [PubMed]

36. Li, J.; McClane, B.A. Contributions of NanI sialidase to Caco-2 cell adherence by *Clostridium perfringens* type A and C strains causing human intestinal disease. *Infect Immun.* **2014**, *82*, 4620–4630. [CrossRef] [PubMed]

37. Chiarezza, M.; Lyras, D.; Pidot, S.J.; Flore-Diaz, M.; Awad, M.M.; Kennedy, C.L.; Cordner, L.M.; Phumoonna, T.; Poon, R.; Hughes, M.L.; et al. The NanI and NanJ sialidases of *Clostridium perfringens* are not essential for virulence. *Infect Immun.* **2009**, *77*, 4421–4428. [CrossRef] [PubMed]

38. Myers, G.S.; Rasko, D.A.; Cheung, J.K.; Ravel, J.; Seshadri, R.; DeBoy, R.T.; Ren, Q.; Varga, J.; Awad, M.M.; Brinkac, L.M.; et al. Skewed genomic variability in strains of the toxigenic bacterial pathogen, *Clostridium perfringens*. *Genome Res.* **2006**, *16*, 1031–1040. [CrossRef] [PubMed]

39. Shimizu, T.; Ohtani, K.; Hirakawa, H.; Ohshima, K.; Yamashita, A.; Shiba, T.; Ogasawara, N.; Hattori, M.; Kuhara, S.; Hayashi, H. Complete genome sequence of *Clostridium perfringens*, an anaerobic flesh-eater. *Proc. Natl. Acad. Sci. USA* **2002**, *99*, 996–1001. [CrossRef] [PubMed]

40. Therit, B.; Cheung, J.K.; Rood, J.I.; Melville, S.B. NanR, a transcriptional regulator that binds to the promoters of genes involved in sialic acid metabolism in the anaerobic pathogen *Clostridium perfringens*. *PLoS ONE* **2015**, *10*, e0133217. [CrossRef] [PubMed]

41. Ohtani, K.; Hirakawa, H.; Tashiro, K.; Yoshizawa, S.; Kuhara, S.; Shimizu, T. Identification of a two-component VirR/VirS regulon in *Clostridium perfringens*. *Anaerobe* **2010**, *16*, 258–264. [CrossRef] [PubMed]

42. Ohtani, K.; Shimizu, T. Regulation of toxin gene expression in *Clostridium perfringens*. *Res. Microbiol.* **2014**, *166*, 280–289. [CrossRef] [PubMed]

43. Hiscox, T.J.; Harrison, P.F.; Chakravorty, A.; Choo, J.M.; Ohtani, K.; Shimizu, T.; Cheung, J.K.; Rood, J.I. Regulation of sialidase production in *Clostridium perfringens* by the orphan sensor histidine kinase ReeS. *PLoS ONE* **2013**, *8*, e73525. [CrossRef] [PubMed]

44. Hiscox, T.J.; Chakravorty, A.; Choo, J.M.; Ohtani, K.; Shimizu, T.; Cheung, J.K.; Rood, J.I. Regulation of virulence by the RevR response regulator in *Clostridium perfringens*. *Infect. Immun.* **2011**, *79*, 2145–2153. [CrossRef] [PubMed]

45. Li, J.; Ma, M.; Sarker, M.R.; McClane, B.A. CodY is a global regulator of virulence-associated properties for *Clostridium perfringens* type D strain CN3718. *mBio* **2013**, *4*, e00770–e00713. [CrossRef] [PubMed]

46. Walters, D.M.; Stirewalt, V.L.; Melville, S.B. Cloning, sequence, and transcriptional regulation of the operon encoding a putative N-acetyl-mannosamine-6-phosphate epimerase (nanE) and sialic acid lyase (nanA) in *Clostridium perfringens*. *J. Bacteriol.* **1999**, *181*, 4526–4532. [PubMed]

47. Olson, M.E.; King, J.M.; Yahr, T.L.; Horswill, A.R. Sialic acid catabolism in *Staphylococcus aureus*. *J. Bacteriol.* **2013**, *195*, 1779–1788. [CrossRef] [PubMed]

48. Almagro-Moreno, S.; Boyd, E.F. Bacterial catabolism of nonulosonic (sialic) acid and fitness in the gut. *Gut Microbes* **2010**, *1*, 45–50. [CrossRef] [PubMed]

49. Almagro-Moreno, S.; Boyd, E.F. Sialic acid catabolism confers a competitive advantage to pathogenic *Vibrio cholerae* in the mouse intestine. *Infect. Immun.* **2009**, *77*, 3807–3816. [CrossRef] [PubMed]

50. Ng, K.M.; Ferreyra, J.A.; Higginbottom, S.K.; Lynch, J.B.; Kashyap, P.C.; Gopinath, S.; Naidu, N.; Choudhury, B.; Weimer, B.C.; Monack, D.M.; et al. Microbiota-liberated host sugars facilitate post-antibiotic expansion of enteric pathogens. *Nature* **2013**, *502*, 96–99. [CrossRef] [PubMed]

51. Ley, R.E. Harnessing microbiota to kill a pathogen: The sweet tooth of *Clostridium difficile*. *Nat. Med.* **2014**, *20*, 248–249. [CrossRef] [PubMed]

52. Brittan, J.; Bucheridge, T.; Finn, A.; Kadioglu, A.; Jenkinson, H. Pneumococcal neuraminidase A: An essential upper airway colonization factor for *Streptococcus pneumoniae*. *Mol. Oral Microb.* **2012**, *27*, 270–283. [CrossRef] [PubMed]

53. Awad, M.M.; Singleton, J.; Lyras, D. The sialidase NanS enhances non-TcsL mediated cytotoxicity of *Clostridium sordellii*. *Toxins* **2016**, *8*. [CrossRef] [PubMed]

54. Chang, D.E.; Smalley, D.J.; Tucker, D.L.; Leatham, M.P.; Norris, W.E.; Stevenson, S.J.; Anderson, A.B.; Grissom, J.E.; Laux, D.C.; Cohen, P.S.; et al. Carbon nutrition of *Escherichia coli* in the mouse intestine. *Proc. Natl. Acad. Sci. USA* **2004**, *101*, 7427–7432. [CrossRef] [PubMed]

55. Galen, J.E.; Ketley, J.M.; Fasano, A.; Richardson, S.H.; Wasserman, S.S.; Kaper, J.B. Role of *Vibrio cholerae* neuraminidase in the fuction of cholera toxin. *Infect. Immun.* **1992**, *60*, 406–415. [PubMed]

56. Jeong, H.G.; Oh, M.H.; Kim, B.S.; Lee, M.Y.; Han, H.J.; Choi, S.H. The capability of catabolic utilization of N-acetylneuraminic acid, a sialic acid, is essential for *Vibrio vulnificus* pathogenesis. *Infect. Immun.* **2009**, *77*, 3209–3217. [CrossRef] [PubMed]

57. Li, J.; Freedman, J.C.; McClane, B.A. NanI sialidase, CcpA, and CodY work together to regulate epsilon toxin production by *Clostridium perfringens* type D strain CN3718. *J. Bacteriol.* **2015**, *197*, 3339–3353. [CrossRef] [PubMed]

58. Shimanoto, S.; Tamai, E.; Matsushita, O.; Minami, J.; Okabe, A.; Miyata, S. Changes in ganglioside content affect the binding of *Clostridium perfringens* epsilon-toxin to detergent-resistant membranes of Madin-Darby canine kidney cells. *Microbiol. Immun.* **2005**, *49*, 245–253. [CrossRef]

59. Nagahama, M.; Sakurai, J. High-affinity binding of *Clostridium perfingens* epsilon-toxin to rat brain. *Infect. Immun.* **1992**, *60*, 1237–1240. [PubMed]

60. Jost, B.H.; Billington, S.J.; Trinh, H.T.; Songer, J.G. Association of genes encoding beta2 toxin and a collagen binding protein in *Clostridium perfringens* isolates of porcine origin. *Vet. Microbiol.* **2006**, *115*, 173–182. [CrossRef] [PubMed]

61. Hitsumoto, Y.; Morita, N.; Yamazoe, R.; Tagomori, M.; Yamasaki, T.; Katayama, S. Adhesive properties of *Clostridium perfringens* to extracellular matrix proteins collagens and fibronectin. *Anaerobe* **2014**, *25*, 67–71. [CrossRef] [PubMed]

62. Katayama, S.; Tagomori, M.; Morita, N.; Yamasaki, T.; Nariya, H.; Okada, M.; Watanabe, M.; Hitsumoto, Y. Determination of the *Clostridium perfringens*-binding site on fibronectin. *Anaerobe* **2015**, *34*, 174–181. [CrossRef] [PubMed]

63. Wade, B.; Keyburn, A.L.; Haring, V.; Ford, M.; Rood, J.I.; Moore, R.J. The adherent abilities of *Clostridium perfringens* strains are critical for the pathogenesis of avian necrotic enteritis. *Vet. Microbiol.* **2016**, *197*, 53–61. [CrossRef]

64. Cioffi, D.L.; Pandey, S.; Alvarez, D.F.; Cioffi, E.A. Terminal sialic acids are an important determinant of pulmonary endothelial barrier integrity. *Am. J. Physiol. Lung Cell. Mol. Physiol.* **2012**, *302*, L1067–1077. [CrossRef] [PubMed]

65. Freedman, J.C.; Li, J.; Uzal, F.A.; McClane, B.A. Proteolytic processing and activation of *Clostridium perfringens* epsilon toxin by caprine small intestinal contents. *mBio* **2014**, *5*, e01994–e01914. [CrossRef] [PubMed]

66. Hanna, P.C.; Wieckowski, E.U.; Mietzner, T.A.; McClane, B.A. Mapping functional regions of *Clostridium perfringens* type A enterotoxin. *Infect. Immun.* **1992**, *60*, 2110–2114. [PubMed]

67. Streicher, H. Inhibition of microbial sialidases-what has happened beyond the influenza virus? *Curr. Med. Chem. Anti-Infect. Agents* **2004**, *3*, 149–161. [CrossRef]

68. Roy, S.; Honma, K.; Douglas, C.W.; Sharma, A.; Stafford, G.P. Role of sialidase in glycoprotein utilization by *Tannerella forsythia*. *Microbiology* **2011**, *157*, 3195–3202. [CrossRef] [PubMed]

69. Holzer, C.T.; von Itzstein, M.; Jin, B.; Pegg, M.S.; Stewart, W.P.; Wu, W.Y. Inhibition of sialidases from viral, bacterial and mammalian sources by analogues of 2-deoxy-2,3-didehydro-*N*-acetylneuraminic acid modified at the C-4 position. *Glycoconj. J.* **1993**, *10*, 40–44. [CrossRef] [PubMed]
70. Knapp, S.; Zhao, D. Synthesis of the sialidase inhibitor siastatin B. *Org. Lett.* **2000**, *2*, 4037–4040. [CrossRef] [PubMed]

Article

Role of p38$_{alpha/beta}$ MAP Kinase in Cell Susceptibility to *Clostridium sordellii* Lethal Toxin and *Clostridium difficile* Toxin B

Ilona Schelle, Janina Bruening, Mareike Buetepage and Harald Genth *

Institute for Toxicology, Hannover Medical School, Carl-Neuberg-Str. 1, D-30625 Hannover, Germany;
ilona.schelle@t-online.de (I.S.); janina.bruening@twincore.de (J.B.); mbuetepage@ukaachen.de (M.B.)
* Correspondence: genth.harald@mh-hannover.de; Tel.: +49-511-532-9168

Academic Editor: Holger Barth
Received: 12 September 2016; Accepted: 19 December 2016; Published: 22 December 2016

Abstract: Lethal Toxin from *Clostridium sordellii* (TcsL), which is casually involved in the toxic shock syndrome and in gas gangrene, enters its target cells by receptor-mediated endocytosis. Inside the cell, TcsL mono-O-glucosylates and thereby inactivates Rac/Cdc42 and Ras subtype GTPases, resulting in actin reorganization and an activation of p38 MAP kinase. While a role of p38 MAP kinase in TcsL-induced cell death is well established, data on a role of p38 MAP kinase in TcsL-induced actin reorganization are not available. In this study, TcsL-induced Rac/Cdc42 glucosylation and actin reorganization are differentially analyzed in p38$_{alpha}^{-/-}$ MSCV empty vector MEFs and the corresponding cell line with reconstituted p38$_{alpha}$ expression (p38$_{alpha}^{-/-}$ MSCV p38$_{alpha}$ MEFs). Genetic deletion of p38$_{alpha}$ results in reduced susceptibility of cells to TcsL-induced Rac/Cdc42 glucosylation and actin reorganization. Furthermore, SB203580, a pyridinyl imidazole inhibitor of p38$_{alpha/beta}$ MAP kinase, also protects cells from TcsL-induced effects in both p38$^{-/-}$ MSCV empty vector MEFs and in p38$_{alpha}^{-/-}$ MSCV p38$_{alpha}$ MEFs, suggesting that inhibition of p38$_{beta}$ contributes to the protective effect of SB203580. In contrast, the effects of the related *C. difficile* Toxin B are responsive neither to SB203580 treatment nor to p38$_{alpha}$ deletion. In conclusion, the protective effects of SB203580 and of p38$_{alpha}$ deletion are likely not based on inhibition of the toxins' glucosyltransferase activity rather than on inhibited endocytic uptake of specifically TcsL into target cells.

Keywords: endocytosis; Ras; *C. difficile* Toxin B; mono-O-glucosylation; p21-activated kinase; actin

1. Introduction

Toxin-producing strains of *C. difficile* and *C. sordellii* cause intestinal infections, including *C. difficile*-associated diarrhea (CDAD) in humans and horses, and *C. sordellii*-induced hemorrhagic enteritis and enterotoxemia in cattle, sheep, and other ruminants [1–4]. The major virulence factors involved in these infections are toxin A (TcdA) and toxin B (TcdB) of *C. difficile*, and lethal toxin (TcsL) and hemorrhagic toxin (TcsH) from *C. sordellii*. These single chained toxins exhibit an AB-like toxin structure with the C-terminal delivery domain mediating cell entry of the N-terminal glucosyltransferase domain by receptor-mediated endocytosis [5,6]. The endocytosed glucosyltransferase domain associates with membrane phosphatidylserine facilitating mono-O-glucosylation of small GTPases of the Rho and Ras subfamilies in a monovalent and divalent metal ion-dependent manner [7–9]. The acceptor amino acid of toxin-catalyzed mono-O-glucosylation is Thr-35 in Rac1, Cdc42, and (H/K/N)Ras and Thr-37 in Rho(A/B/C). Mono-O-glucosylated Rho/Ras GTPases are incapable of coupling to their regulatory and effector protein and thus are functionally inactive [10–13]. Treatment of cultured cells with the glucosylating toxins results in a breakdown of the actin-based cytoskeleton ("cytopathic effect") and (at higher toxin concentrations) in cell death ("cytotoxic effect") based on inhibition of Rho/Ras-dependent signaling pathways regulating actin dynamics [14], cell-matrix binding [15,16], cell cycle progression, and cell survival [17–19].

The family of p38 MAPKs encompasses the four isoforms $p38_{alpha}$, $p38_{beta}$, $p38_{gamma}$, and $p38_{delta}$. $p38_{alpha}$ and $p38_{beta}$ are ubiquitously expressed, while $p38_{gamma}$ and $p38_{delta}$ exhibit a more restricted expression patterns. The best characterized isoform is $p38_{alpha}$, which is involved in the regulation of the main cellular functions, including actin dynamics, differentiation, and cell death and survival [20,21]. TcsL (as well as TcdA and TcdB) has been shown to activate the family of mitogen-activated protein kinases (MAPKs) involving ERKs (extracellular signal-regulated kinases), JNKs (c-Jun-*N*-terminal kinases) and p38 MAPKs [22]. MAP kinase signaling cascades have been shown to be involved in the cytopathic as well as the cytotoxic effects of the glucosylating toxins. In particular, TcsL-induced activation of JNK has been suggested to facilitate TcsL-catalyzed GTPase substrate glucosylation and subsequently TcsL-induced actin reorganization [22]. $p38_{alpha}$ MAP kinase signaling has been implicated in TcsL-induced expression of the cell death-regulating GTPase RhoB [23]. Finally, TcsL-induced cell death has been shown to be responsive to inhibition by SB203580, a pyridinyl imidazole inhibitor of $p38_{alpha/beta}$ MAP kinase, suggesting a role of $p38_{alpha/beta}$ in TcsL-induced cell death [24].

In this study, TcsL-induced Rac/Cdc42 glucosylation and subsequent actin reorganization were analyzed for an involvement of $p38_{alpha}$ in murine embryonic fibroblasts (MEFs). In this study, TcsL-induced Rac/Cdc42 glucosylation and actin reorganization are differentially analyzed in $p38_{alpha}^{-/-}$ murine stem cell virus empty vector MEFs ($p38_{alpha}^{-/-}$ MSCV EV MEFs) and the corresponding cell line with reconstituted $p38_{alpha}$ expression ($p38_{alpha}^{-/-}$ MSCV $p38_{alpha}$ MEFs) [25,26]. We show that genetic deletion of $p38_{alpha}$ as well as SB203580 protects cells from TcsL-catalyzed glucosylation of Rac/Cdc42 subtype GTPases and TcsL-induced actin reorganization.

2. Results

2.1. Prevention of TcsL-Induced Actin Re-Organization upon Inhibition of $p38_{alpha/beta}$

TcsL time-dependently induced actin reorganization in $p38_{alpha}$-proficient $p38_{alpha}^{-/-}$ MSCV $p38_{alpha}$ MEFs (Figure 1), with TcsL treatment for 4 h being sufficient for almost complete cell rounding (Figure 1). In contrast, TcsL-induced rounding of $p38_{alpha}$-deficient $p38_{alpha}^{-/-}$ MSCV EV MEFs was clearly delayed, suggesting a reduced susceptibility of $p38_{alpha}^{-/-}$ MSCV EV MEFs to TcsL. Next, TcsL-induced cell rounding was analyzed in $p38_{alpha}^{-/-}$ MSCV $p38_{alpha}$ MEFs treated with SB203580, a pyridinyl imidazole inhibitor of $p38_{alpha/beta}$ MAP kinase [27]. SB203080 concentration-dependently reduced TcsL-induced cell rounding, with a SB203580 concentration of 10 µM being sufficient for almost complete prevention of TcsL-induced cell rounding (Figure 2A,B). A pronounced protective effect of SB203580 was also observed in a time-dependent experiment (Figure 2C). SB203580 (10 µM) alone did not change fibroblast morphology (Figure 2A). Finally, the protective effect of $p38_{alpha}$ inhibition was analyzed in TcsL concentration-dependent experiments (Figure 3). Either genetic deletion of $p38_{alpha}$ or SB203580 treatment delayed TcsL-induced cell rounding. Interestingly, SB203580 treatment of $p38_{alpha}$-deficient $p38_{alpha}^{-/-}$ MSCV EV MEFs further delayed TcsL-induced cell rounding, suggesting a role of $p38_{beta}$ inhibition in the reduced susceptibility to TcsL.

TcdB is highly related to TcsL (identity of 75% at amino acid level), as both TcdB and TcsL enter their target cells by receptor-mediated endocytosis and both cause changed actin dynamics by mono-O-glucosylation of small GTPases. Interestingly, neither genetic deletion (Figure 4A,B) nor treatment with the $p38_{alpha/beta}$ inhibitor SB203580 (Figure 4C,D) changed the kinetics of TcdB-induced actin reorganization. $p38_{alpha/beta}$ inhibition thus mediates protection of fibroblasts from TcsL-(not TcdB-) induced actin reorganization.

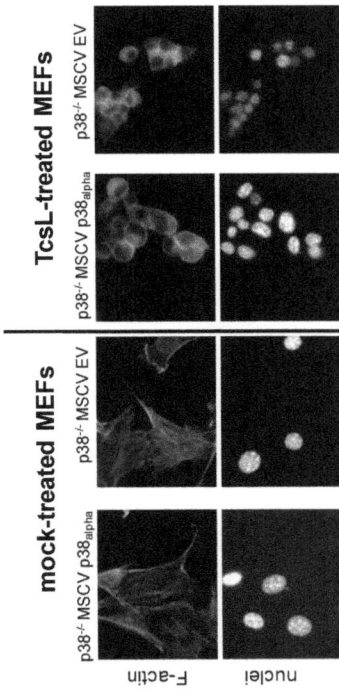

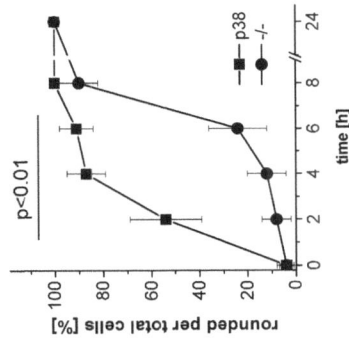

Figure 1. Effects of genetic deletion of $p38_{alpha}$ on TcsL-induced changes of cell morphology (time-dependency). $p38_{alpha}^{-/-}$ MSCV empty vector (EV) MEFs and the corresponding cell line with reconstituted $p38_{alpha}$ expression ($p38_{alpha}^{-/-}$ MSCV $p38_{alpha}$ MEFs) were treated with TcsL (1 µg/mL) for the indicated times. Cells were then washed, fixed, permeabilized, and stained with rhodamine-phalloidin and DAPI. Cell morphology was visualized using fluorescence microscopy (20× amplification). TcsL-induced changes of the morphology were time-dependently quantified in terms of the number of rounded per total cells. Six representative microscopic fields were chosen and 300 cells total were counted for characteristic cell rounding. Values are the mean ± SD from three independent experiments performed in triplicates. $p < 0.01$ indicates significant differences comparing $p38_{alpha}$-proficient with $p38_{alpha}$-deficient cells using Student's *t*-test.

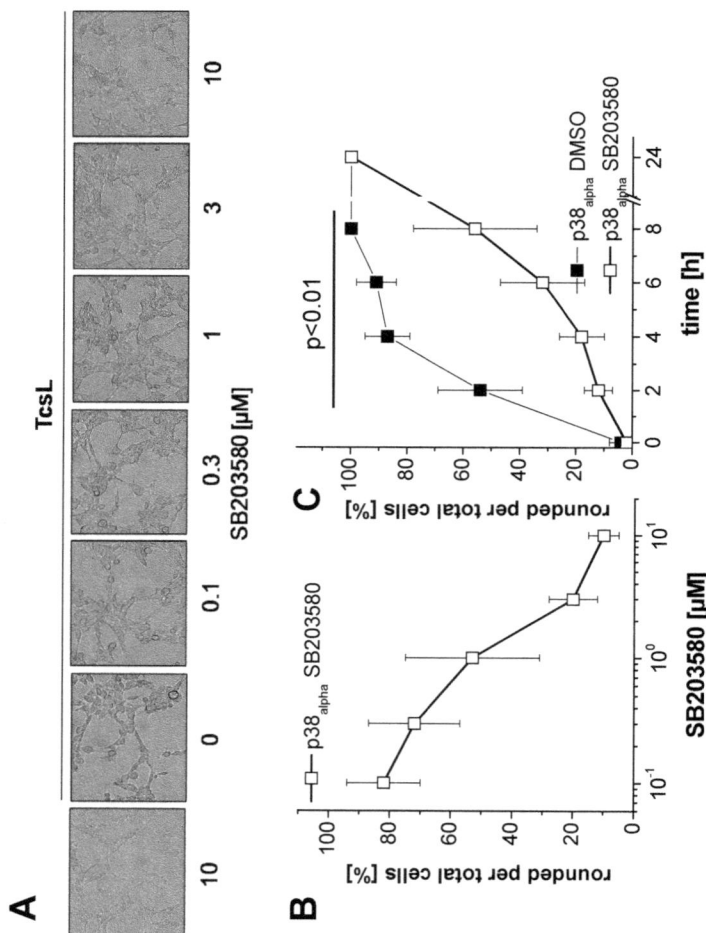

Figure 2. Effects of SB203580 treatment on TcsL-induced changes of cell morphology: (**A**) p38$_{alpha}$$^{-/-}$ MSCV p38$_{alpha}$ MEFs were treated with TcsL (1 μg/mL) or buffer in the presence of the indicated concentrations of SB203580 for 4 h. Cell morphology was visualized using phase contrast microscopy (10× amplification). (**B**) TcsL-induced changes of the morphology were quantified in terms of the number of rounded per total cells. Six representative microscopic fields were chosen and 300 cells total were counted for characteristic cell rounding. Values are the mean ± SD from three independent experiments performed in triplicates. (**C**) p38$_{alpha}$$^{-/-}$ MSCV p38$_{alpha}$ MEFs were treated with TcsL (1 μg/mL) in the presence of SB203580 (10 μM) or DMSO alone for the indicated times. TcsL-induced changes of the morphology were quantified in terms of the number of rounded per total cells. Values are the mean ± SD from three independent experiments performed in triplicates. *p* < 0.01 indicates significant differences comparing SB203580-treated with DMSO-treated cells using Student's *t*-test.

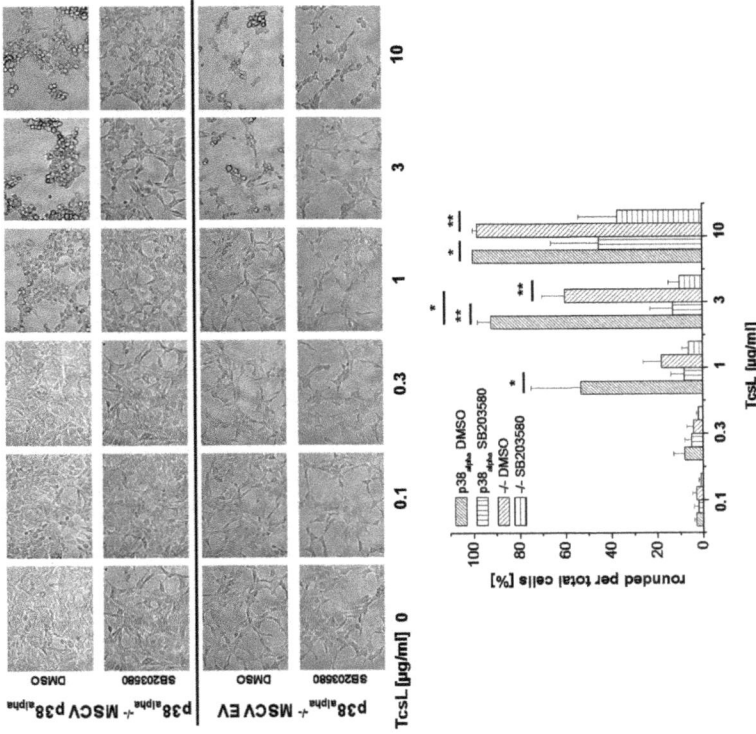

Figure 3. Effects of genetic deletion of p38$_{alpha}$ and of SB203580 treatment on TcsL-induced changes of cell morphology (concentration-dependency). p38$_{alpha}$$^{-/-}$ MSCV p38$_{alpha}$ MEFs and p38$_{alpha}$$^{-/-}$ MSCV empty vector (EV) MEFs were treated with the indicated concentrations of TcsL in the presence of SB203580 (10 μM) or DMSO alone for 4 h. Cell morphology was visualized using phase contrast microscopy (10× amplification). TcsL-induced changes of the morphology of p38$^{-/-}$ MSCV p38$_{alpha}$ MEFs and of p38$^{-/-}$ MSCV empty vector MEFs were quantified in terms of the number of rounded per total cells. Six representative microscopic fields were chosen and 300 cells total were counted for characteristic cell rounding. Values are the mean ± SD from three independent experiments performed in triplicates. */** indicates significant differences, $p < 0.05$/$p < 0.005$ as analyzed using Student's *t*-test.

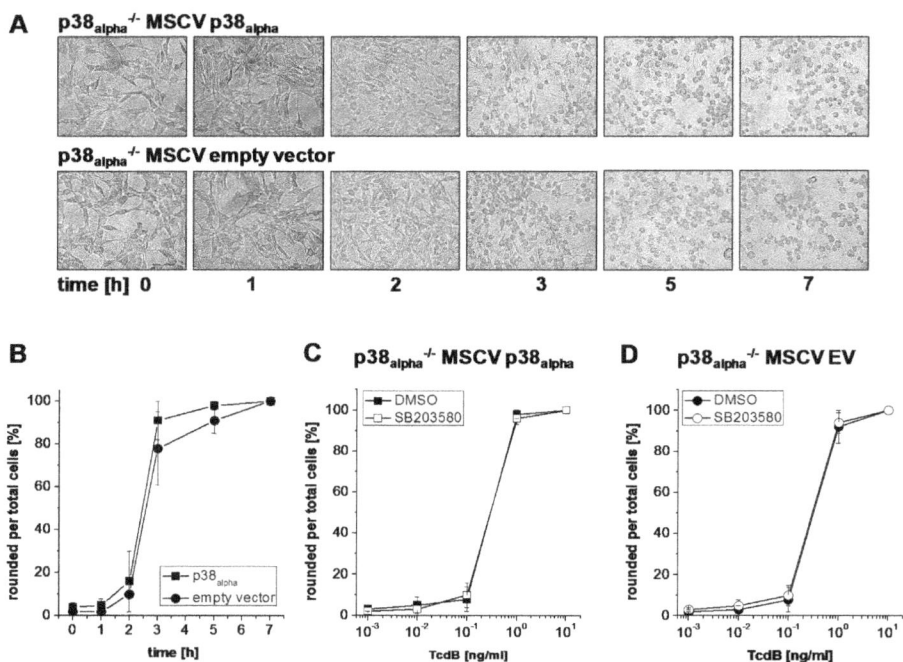

Figure 4. Effects of genetic deletion of p38$_{alpha}$ and of SB203580 treatment on TcdB-induced changes of cell morphology: (**A,B**) p38$_{alpha}^{-/-}$ MSCV p38$_{alpha}$ MEFs and p38$_{alpha}^{-/-}$ MSCV empty vector (EV) MEFs were treated with TcdB (1 ng/mL) for the indicated times; and (**C,D**) p38$_{alpha}^{-/-}$ MSCV p38$_{alpha}$ MEFs and p38$_{alpha}^{-/-}$ MSCV empty vector MEFs were treated with the indicated concentrations of TcdB in the presence of SB203580 (10 µM) or DMSO alone for 4 h. Cell morphology was visualized using phase contrast microscopy (10× amplification). TcdB-induced changes of the morphology were quantified in terms of the number of rounded per total cells. Six representative microscopic fields were chosen and 300 cells total were counted for characteristic cell rounding. Values are the mean ± SD from three independent experiments performed in triplicates.

2.2. Prevention of TcsL-Induced Glucosylation of Rac/Cdc42 Subtype GTPases upon Inhibition of p38$_{alpha/beta}$

Actin reorganization has been attributed to TcsL-/TcdB-catalyzed glucosylation of Rac/Cdc42 subtype GTPases and the subsequent loss of cell-matrix binding [15,28,29]. For the analysis of TcsL-catalyzed glucosylation of Rac/Cdc42, lysates from p38$_{alpha}^{-/-}$ MSCV p38$_{alpha}$ MEFs were analyzed by immunoblot analysis using the Rac1(clone 102) antibody. This antibody is specific for non-glucosylated Rac/Cdc42 subtype GTPases [30,31]. Once Rac/Cdc42 subtype GTPases are glucosylated by TcsL, the antibody does not detect its epitope, resulting in a lost signal. TcsL-treated p38$_{alpha}^{-/-}$ MSCV p38$_{alpha}$ MEFs exhibited time-dependent glucosylation of Rac/Cdc42 subtype GTPases (Figure 5A). The cellular level of Rac1 was not changed upon TcsL treatment (as analyzed using the Rac1(clone 23A8) antibody) (Figure 5A), confirming that decreasing detection of Rac/Cdc42 subtype GTPases by the Rac1(Mab 102) antibody was due to glucosylation but not due to degradation. TcsL-catalyzed Rac/Cdc42 glucosylation results in dephosphorylation of the FA component p21-associated kinase1/2 (PAK1/2), which is a Rac/Cdc42 effector protein [15,16]. TcsL treatment time-dependently resulted in decreasing levels of pS144/141-PAK1/2 (Figure 5A), indicating PAK1/2 deactivation [32]. The levels of total PAK2 also time-dependently decreased in TcsL-treated cells (Figure 5), showing that PAK1/2 deactivation was based on both dephosphorylation and degradation. TcsL-catalyzed Rac/Cdc42 glucosylation (i.e., inactivation) was thus reflected by deactivation of its effector kinase PAK1/2. Finally, p38 MAP kinase activity was tracked in terms of phosphorylation of its downstream target MAPKAPK2. TcsL treatment resulted in a transient increase of pT222-MAPKAPK2, indicating transient MAPKAPK2 activation. In p38$_{alpha}$-deficient p38$^{-/-}$ MSCV empty vector MEFs, in which MAPKAPK2 activation was hardly observed, Rac/Cdc42 glucosylation was clearly delayed as compared with p38$_{alpha}$$^{-/-}$ MSCV p38$_{alpha}$ MEFs (Figure 5). In a TcsL concentration-dependent experiment, TcsL turned out to be at least three fold more potent in inducing Rac/Cdc42 glucosylation (Figure 6) and PAK1/2 deactivation (Figure 6) in p38$^{-/-}$ MSCV p38$_{alpha}$ MEFs as compared with p38$^{-/-}$ MSCV empty vector MEFs. In contrast to TcsL, the kinetics of TcdB-induced Rac/Cdc42 glucosylation and of PAK1/2 deactivation were comparable in p38$_{alpha}$$^{-/-}$ MSCV p38$_{alpha}$ and p38$^{-/-}$ MSCV empty vector MEFs (Figure 7). TcdB-catalyzed Rac/Cdc42 glucosylation was thus not susceptible to genetic p38$_{alpha}$ inhibition (Figure 7), consistent with above observations on TcdB-induced actin reorganization (Figure 4). These observations show that genetic p38$_{alpha}$ inhibition protects fibroblasts specifically from TcsL-(not TcdB-) induced cytopathic effects.

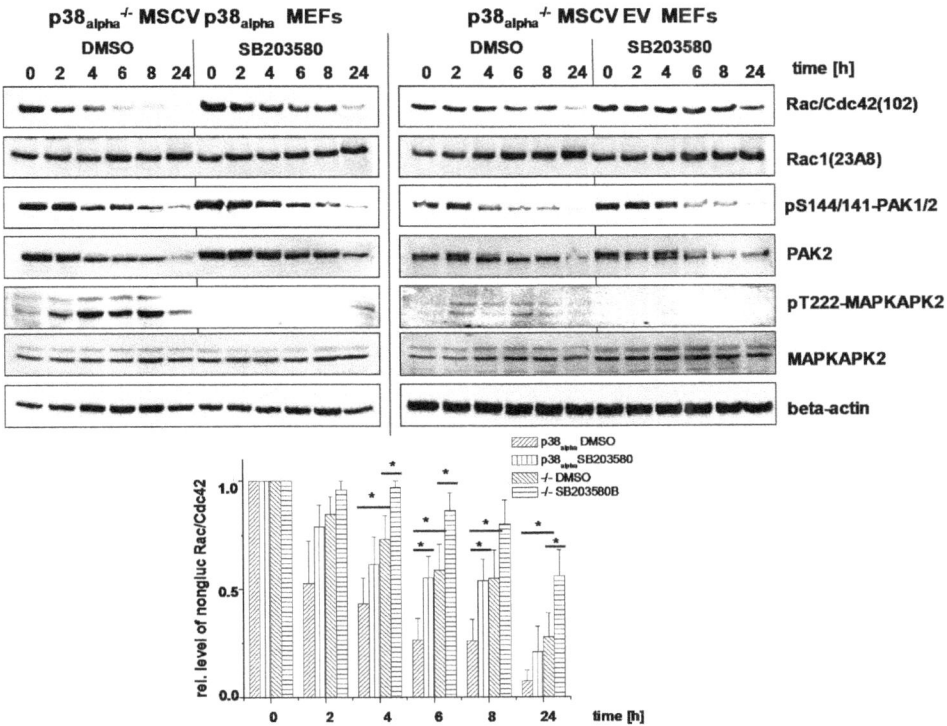

Figure 5. Effects of genetic deletion of p38$_{alpha}$ and of SB203580 treatment on TcsL-catalyzed Rac/Cdc42 glucosylation (time-dependency). p38$_{alpha}$$^{-/-}$ MSCV p38$_{alpha}$ MEFs and p38$_{alpha}$$^{-/-}$ MSCV empty vector (EV) MEFs were treated with TcsL (1 μg/mL) in the presence of SB203580 (10 μM) or DMSO alone for the indicated times. The cellular levels of non-glucosylated Rac/Cdc42, total Rac1, pS144/141-PAK1/2, PAK2, pT222-MAPKAPK2, MAPKAPK2, and beta-actin were analyzed by immunoblotting using the indicated antibodies. Quantifications of immunoblots were performed using Kodak software and relative amounts of non-glucosylated Rac/Cdc42 versus the total levels of Rac1, respectively, are expressed as mean ± SD of three independent experiments. * indicates significant differences, *p* < 0.05, as analyzed using Student's *t*-test.

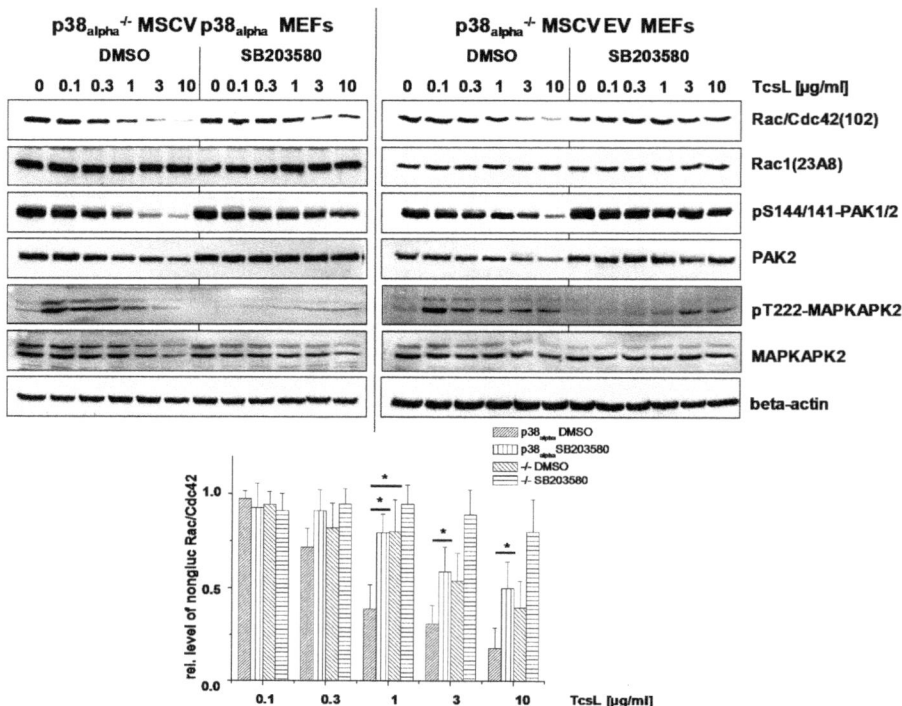

Figure 6. Effects of genetic deletion of p38$_{alpha}$ and of SB203580 treatment on TcsL-catalyzed Rac/Cdc42 glucosylation (TcsL concentration-dependency). p38$_{alpha}$$^{-/-}$ MSCV p38$_{alpha}$ MEFs and p38$_{alpha}$$^{-/-}$ MSCV empty vector (EV) MEFs were treated with the indicated concentrations of TcsL in the presence of SB203580 (10 μM) or DMSO alone for 4 h. The cellular levels of non-glucosylated Rac/Cdc42, total Rac1, pS144/141-PAK1/2, PAK2, and beta-actin were analyzed by immunoblotting using the indicated antibodies. Quantifications of immunoblots were performed using Kodak software and relative amounts of non-glucosylated Rac/Cdc42 versus the total levels of Rac1, respectively, are expressed as mean ± SD of three independent experiments. * indicates significant differences, $p < 0.05$, as analyzed using Student's *t*-test.

TcsL-induced Rac/Cdc42 glucosylation were next analyzed upon pharmacological inhibition of p38$_{alpha/beta}$. SB203580 treatment of either p38$_{alpha}$$^{-/-}$ MSCV p38$_{alpha}$ and p38$_{alpha}$$^{-/-}$ MSCV empty vector MEFs resulted in an almost complete loss of pT222-MAPKAPK2, confirming effective p38 inhibition (Figures 5 and 6). TcsL-catalyzed Rac/Cdc42 glucosylation and PAK1/2 deactivation were responsive to SB203580 treatment in both p38$_{alpha}$$^{-/-}$ MSCV p38$_{alpha}$ and p38$_{alpha}$$^{-/-}$ MSCV empty vector MEFs, both in time- (Figure 5) and concentration-dependent (Figure 6) experiments. The observation that cell rounding and Rac/Cdc42 glucosylation in p38$_{alpha}$$^{-/-}$ MSCV empty vector MEFs are responsive to SB203580 suggests that the protective effects of SB203580 involve inhibition of both p38$_{alpha}$ and p38$_{beta}$. Taken together, p38$_{alpha/beta}$ inhibition-mediated protection of fibroblasts from TcsL-induced actin reorganization coincides with protection from TcsL-catalyzed Rac/Cdc42 glucosylation.

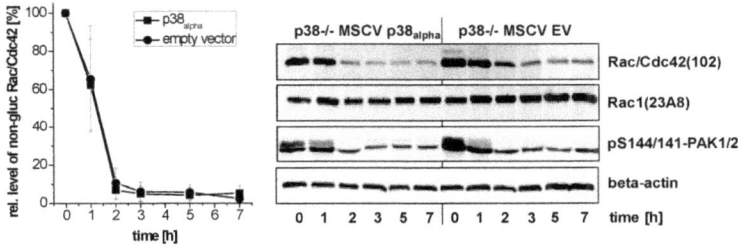

Figure 7. Effects of genetic deletion of p38alpha on TcdB-catalyzed Rac/Cdc42 glucosylation (time-dependency). p38$_{alpha}^{-/-}$ MSCV p38$_{alpha}$ MEFs and p38$_{alpha}^{-/-}$ MSCV empty vector (EV) MEFs were treated with TcdB (1 ng/mL) for the indicated times. The cellular levels of non-glucosylated Rac/Cdc42, total Rac1, pS144/141-PAK1/2, and beta-actin were analyzed by immunoblotting using the indicated antibodies. Quantifications of immunoblots were performed using Kodak software and relative amounts of non-glucosylated Rac/Cdc42 versus the total levels of Rac1, respectively, are expressed as mean ± SD of three experiments.

2.3. SB203580 Preserves TcsL-Induced Loss of Epithelial Barrier Function

C. sordellii-associated disease include necrotic and hemorrhagic enteritis, whereby TcsL has been shown to alter epithelial permeability [33,34]. To check if SB203580-mediated inhibition of TcsL might be useful with regard to disease treatment, TcsL-induced loss of epithelial barrier function was next analyzed in terms of the loss of the transepithelial resistance (TER) of a Madin-Darby canine kidney (MDCK-C7) monolayer [9,35]. TcsL treatment time-dependently decreased the TER of the MDCK-C7 monolayer (Figure 8A). In the presence of SB203580, TcsL-induced loss of TER was markedly attenuated (Figure 8A). SB203580 alone did not affect the TER (Figure 8A). In contrast, TcdB-induced loss of the TER of the MDCK-C7 monolayer was not responsive to SB203580 treatment (Figure 8B), consistent with above observations showing that TcdB-induced actin reorganization was not responsive to SB203580 treatment (Figure 4). SB203580 treatment thus might be useful in the light of treatment of the TcsL-induced loss of epithelial barrier function.

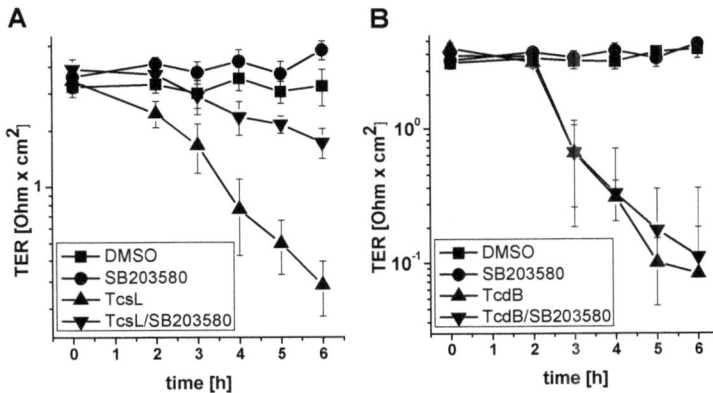

Figure 8. SB203580 preserves TcsL-induced loss of epithelial barrier function. Madin-Darby canine kidney (MDCK-C7) monolayers grown on Transwell filter inserts were treated with TcsL (30 µg/mL (**A**)) and TcdB (30 ng/mL (**B**)) in the presence of SB203580 (10 µM) or DMSO alone as indicated. Transepithelial electrical resistance (TER) was monitored for the indicated times. TER values are given as means ± SD of three independent experiments.

3. Discussion

In this study, SB203580, a pyridinyl imidazole inhibitor of p38$_{alpha/beta}$ MAP kinase, has been presented to efficaciously prevent TcsL-induced loss of epithelial barrier function of MDCK-C7 monolayers and to prevent TcsL-induced cell rounding, Rac/Cdc42 glucosylation, and PAK deactivation in murine fibroblasts. Furthermore, genetic deletion of p38$_{alpha}$ is also sufficient for preventing TcsL-induced cell rounding, Rac/Cdc42 glucosylation, and PAK deactivation in murine fibroblasts. Interestingly, p38$_{alpha}^{-/-}$ MSCV empty vector MEFs turn out to be still sensitive to SB203580, suggesting that the protective effects of SB203580 involves inhibition of both p38$_{alpha}$ and p38$_{beta}$. How does p38 inhibition mediate the protective effect against TcsL?

TcdB and TcsL both are mono-O-glucosyltransferases that modify small GTPases. Their N-terminal glucosyltransferase domains are structurally and functionally highly related, as they share Rac/Cdc42 as substrate GTPases (with threonine-35 being the acceptor amino acid) and UDP-glucose as a sugar donor [36]. From the observation that TcdB-induced Rac/Cdc42 glucosylation is insensitive to p38 inhibition, it can be concluded that SB203580 does not interfere with the intracellular glucosyltransferase activity of the toxins. This leads to the new hypothesis that p38$_{alpha/beta}$ inhibition affects endocytic uptake of TcsL into target cells. In fact, members of the p38 MAP kinase family are important regulators of endocytosis, as they control endocytic trafficking via the GDI-Rab5 complex [37]. In particular, p38 MAP kinase regulates stress-induced internalization of the epidermal growth factor receptor (EGFR) and μ opioid receptor endocytosis [38–40]. Against this background, inhibition of p38$_{alpha/beta}$ might prevent cell entry of specifically TcsL (not TcdB) by receptor-mediated endocytosis. This hypothetic model implies that TcsL and TcdB enter their target cells by exploiting distinct cell surface receptors, with the TcsL cell surface receptors being internalized in a p38$_{alpha/beta}$-dependent and the TcdB cell surface receptors being internalized in a p38$_{alpha/beta}$-independent manner.

In a former study, the responsiveness of TcsL-induced effects such as apoptotic cell death to inhibition by SB203580 has been interpreted in terms of an involvement of p38$_{alpha/beta}$ in TcsL-induced cell death [24]. The observations of this study have led to the conclusion that the protective effect of SB203580 is based on rather inhibition of TcsL uptake than on a role of p38 in the cytopathic effects of TcsL. Against this background, the responsiveness of TcsL-induced cell death to inhibition by SB203580 must be re-interpreted in terms of reduced TcsL uptake and subsequently reduced GTPase substrate glucosylation as the cause of cell death inhibition. In autophagy research, the pyridinyl imidazole inhibitors SB203580 and SB202190 have been shown to interfere with the autophagic flux independently of p38 MAP kinase [41]. These observations have led to the recommendation that pyridinyl imidazole class inhibitors should not be used as pharmacological tools in the analysis of MAPK11-MAPK14/p38-dependence [42]. The latter recommendation seems to be applicable also in the field of protein toxins.

The protective effect of the pyridinyl imidazole SB203580 is most interesting with regard to the development of non-antibiotic treatment for diseases caused by toxigenic *C. sordellii*. Further research will address the characterization of those pathways mediating the protective effect against TcsL upon inhibition of p38$_{alpha/beta}$. Furthermore, a screening of additional pyridinyl imidazole compounds capable of inhibiting the effects of TcsL is under way in our laboratory.

4. Conclusions

- Genetic deletion of p38$_{alpha}$ or treatment with SB203580 protects MEFs from Rac/Cdc42 glucosylation and actin reorganization induced by TcsL (not by the related TcdB).
- Treatment with SB203580 protects epithelial monolayer from loss of epithelial barrier function induced by TcsL (not by TcdB).
- The protective effects of SB203580 treatment and of p38$_{alpha}$ deletion are likely based on inhibition of endocytic uptake of TcsL rather than on inhibition of the toxins' glucosyltransferase activity.

5. Materials and Methods

5.1. Materials

The following reagents were obtained from commercial sources: SB203580 (4-(4-fluorophenyl)-2-(4-methylsulfinylphenyl)-5-(4-pyridyl)imidazole) (Calbiochem, Darmstadt, Germany); DAPI (40.6-diamidino-2-phenylindole) (Sigma-Aldrich, St. Louis, MO, USA); and rhodamine-conjugated phalloidin (Sigma-Aldrich, St. Louis, MO, USA).

Toxins: TcsL was prepared from *C. sordellii* IP82, which is the same strain as 6018, and TcdB from *C. difficile* VPI10463. Toxins were produced and purified yielding only one band on SDS-PAGE as previously described [43,44]. In brief, a dialysis bag containing 900 mL of 0.9% NaCl in a total volume of 4 liters of brain heart infusion (Difco, BD Life Sciences, Heidelberg, Germany) was inoculated with 100 mL of an overnight culture of *C. sordellii* or *C. difficile*. The culture was grown under microaerophilic conditions at 37 °C for 72 h. Bacteria were removed from the dialysis bag solution by centrifugation. Proteins from the culture supernatant from were precipitated by ammonium sulfate (Merck Millipore, Darmstadt, Germany) at 70% saturation. The precipitated proteins were dissolved in 50 mM Tris-HCl pH 7.5 buffer and extensively dialyzed against 50 mM Tris-HCl pH 7.5 buffer for 24 h. The protein solution was loaded onto an anion exchange column (MonoQ, GE Healthcare Europe, Freiburg, Germany). Either TcsL or TcdB were eluted with 50 mM Tris-HCl, pH 7.5, at 500–600 mM NaCl and were subsequently dialyzed against buffer (50 mM Tris-HCl pH 7.5, 15 mM NaCl). The absence of TcdA (which eluted at 150–200 mM NaCl) in TcdB preparations was checked by immunoblot analysis.

5.2. Cell Culture and Preparation of Lysates

$p38^{-/-}$ MSCV empty vector MEFs and the corresponding $p38^{-/-}$ MSCV $p38_{alpha}$ MEFs (kindly provided by Dr. Angel Nebreda, Institute for Research in Biomedicine, Barcelona, Spain) were cultivated in Dulbecco's modified essential medium supplemented with 10% FCS, 100 µg/mL penicillin, 100 U/mL streptomycin and 1 mM sodium pyruvate at 37 °C and 5% CO_2 according to standard protocols [45]. Cells sub-confluently seeded in 3.5-cm dishes were treated with TcsL, TcdB, and SB203580 for different times and concentrations as noted in the figures. Thereby, cells were pretreated with 10 µM SB202580 dissolved in DMSO (final DMSO concentration in the medium 2%) for 20 min and subsequently treated with the toxins or buffer. Upon incubation time, the cells were rinsed with 5 mL of ice-cold phosphate-buffered saline and scraped off in 200 µL of Laemmli lysis buffer per dish. The cells were disrupted mechanically by sonification (five times on ice). The lysate were submitted to immunoblot analysis.

5.3. Immunoblot Analysis

Cells lysates were separated on 15% polyacrylamide gels and transferred onto nitrocellulose for 2 h at 250 mA, followed by blocking with 5% (*w/v*) nonfat dried milk for 1 h. Blots were incubated with the appropriate primary antibody with dilution according to the manufacturers' instructions (beta-actin, Mab AC-40, Sigma-Aldrich, St. Louis, MO, USA; dilution 1:5000); MAPKAPK-2 (Cell signaling 3042, dilution 1:1000); pT222-MAPKAPK-2 (Cell signaling 3316, dilution 1:1000); PAK2 (Cell signaling 2608, dilution 1:1000); phospho-S144/141-PAK1/2 (Abcam ab40795; dilution 1:2000); Rac1 (BD Transduction Laboratories 610650, clone 102; dilution 1:1000); Rac1(Millipore 05-389, clone 23A8; dilution 1:1000) in buffer B (50 mM Tris-HCl, pH 7.2, 150 mM NaCl, 5 mM KCl, 0.05% (*w/v*) Tween 20) for 18 h and subsequently for 2 h with a horseradish peroxidase-conjugated secondary antibody (mouse: Rockland 610-1034-121; dilution 1:3000; rabbit Rockland 611-1302; dilution 1:3000). For the chemiluminescence reaction, ECL Femto (Fisher Scientific, Schwerte, Germany) was used. The signals were analyzed densitometrically using the KODAK 1D software (2004, Rochester, MN, USA).

5.4. Visualization of the Actin Cytoskeleton

p38$_{alpha}^{-/-}$ MSCV empty vector MEFs and the corresponding p38$_{alpha}^{-/-}$ MSCV p38$_{alpha}$ MEFs were grown on coverslips, fixed with formaldehyde (4%) in PBS and permeabilized with 0.1% Triton X-100 (Sigma-Aldrich, St. Louis, MO, USA). The actin cytoskeleton and the nuclei were labeled using rhodamine-phalloidin and DAPI, respectively. Fluorescence images were recorded using a Zeiss Axiovert M (Jena, Germany).

5.5. Transepithelial Resistance of Epithelial Monolayers

Madin-Darby canine kidney (MDCK-C7) cells were cultured under standard conditions (37 °C, 5% CO$_2$) as described [35]. Briefly, MDCK-C7 cells were cultured in minimum essential medium (MEM) enriched with Earle's salts, non-essential amino acids, glutamic acid and 10% fetal calf serum (Biochrom, Berlin, Germany) and split twice weekly using standard culture techniques. MDCK-C7 cells were seeded onto 12 well filter transwell inserts (pore size 0.4 μM, BD Life Sciences, Heidelberg, Germany). The transepithelial electrical resistance (TER) was determined by a Voltohmmeter equipped with Endom 24 chamber (EVOM, World Precision Instruments, Berlin, Germany). MDCK-C7 monolayers were cultivated up to an initial resistance of >2 kΩ·cm^2. The toxins and SB203580 (final DMSO concentration in the medium 2%) were applied on the basolateral site of the monolayer and toxin-induced loss of TER was analyzed in a time-dependent manner.

Acknowledgments: p38$^{-/-}$ MSCV empty vector MEFs and the corresponding p38$^{-/-}$ MSCV p38$_{alpha}$ MEFs were kindly provided by Angel Nebreda, Institute for Research in Biomedicine, Barcelona, Spain. This work was supported by Niedersachsen Vorab "Epidemiology and systems biology of the bacterial pathogen *Clostridium difficile*" grant number VWZN3215 to H.G.

Author Contributions: I.S., J.B., M.B., and H.G. conceived and designed the experiments and analyzed the data; I.S., M.B., and J.B. performed the experiments; and H.G. wrote the paper.

Conflicts of Interest: The authors declare no conflict of interest. The founding sponsors had no role in the design of the study; in the collection, analyses, or interpretation of data; in the writing of the manuscript, and in the decision to publish the results.

References

1. Hodges, K.; Hecht, G. Bacterial infections of the small intestine. *Curr. Opin. Gastroenterol.* **2013**, *29*, 159–163. [CrossRef] [PubMed]
2. Peniche, A.G.; Savidge, T.C.; Dann, S.M. Recent insights into *Clostridium difficile* pathogenesis. *Curr. Opin. Infect. Dis.* **2013**, *26*, 447–453. [CrossRef] [PubMed]
3. Diab, S.S.; Songer, G.; Uzal, F.A. *Clostridium difficile* infection in horses: A review. *Vet. Microbiol.* **2013**, *167*, 42–49. [CrossRef] [PubMed]
4. Al-Mashat, R.R.; Taylor, D.J. Production OD diarrhoea and enteric lesions in calves by the oral inoculation of pure cultures of *Clostridium sordellii*. *Vet. Rec.* **1983**, *112*, 141–146. [CrossRef] [PubMed]
5. Papatheodorou, P.; Zamboglou, C.; Genisyuerek, S.; Guttenberg, G.; Aktories, K. Clostridial glucosylating toxins enter cells via clathrin-mediated endocytosis. *PLoS ONE* **2010**, *5*, e10673. [CrossRef] [PubMed]
6. Jank, T.; Belyi, Y.; Aktories, K. Bacterial glycosyltransferase toxins. *Cell. Microbiol.* **2015**, *17*, 1752–1765. [CrossRef] [PubMed]
7. Varela, C.C.; Haustant, G.M.; Baron, B.; England, P.; Chenal, A.; Pauillac, S.; Blondel, A.; Popoff, M.R. The Tip of the four N-Terminal alpha-helices of *Clostridium sordellii* lethal toxin contains the interaction site with membrane phosphatidylserine facilitating small GTPases glucosylation. *Toxins (Basel)* **2016**, *8*, E90. [CrossRef] [PubMed]
8. LaFrance, M.E.; Farrow, M.A.; Chandrasekaran, R.; Sheng, J.; Rubin, D.H.; Lacy, D.B. Identification of an epithelial cell receptor responsible for *Clostridium difficile* TcdB-induced cytotoxicity. *Proc. Natl. Acad. Sci. USA* **2015**, *112*, 7073–7078. [CrossRef] [PubMed]
9. Genth, H.; Schelle, I.; Just, I. Metal Ion Activation of *Clostridium sordellii* Lethal Toxin and *Clostridium difficile* Toxin B. *Toxins (Basel)* **2016**, *8*, E109. [CrossRef] [PubMed]

10. Sehr, P.; Joseph, G.; Genth, H.; Just, I.; Pick, E.; Aktories, K. Glucosylation and ADP-ribosylation of Rho proteins—Effects on nucleotide binding, GTPase activity, and effector-coupling. *Biochemistry* **1998**, *37*, 5296–5304. [CrossRef] [PubMed]

11. Genth, H.; Aktories, K.; Just, I. Monoglucosylation of RhoA at Threonine-37 blocks cytosol-membrane cycling. *J. Biol. Chem.* **1999**, *274*, 29050–29056. [CrossRef] [PubMed]

12. Herrmann, C.; Ahmadian, M.R.; Hofmann, F.; Just, I. Functional consequences of monoglucosylation of H-Ras at effector domain amino acid threonine-35. *J. Biol. Chem.* **1998**, *273*, 16134–16139. [CrossRef] [PubMed]

13. Vetter, I.R.; Hofmann, F.; Wohlgemuth, S.; Herrmann, C.; Just, I. Structural consequences of mono-glucosylation of Ha-Ras by *Clostridium sordellii* lethal toxin. *J. Mol. Biol.* **2000**, *301*, 1091–1095. [CrossRef] [PubMed]

14. May, M.; Wang, T.; Muller, M.; Genth, H. Difference in F-actin depolymerization induced by toxin B from the *Clostridium difficile* strain VPI 10463 and toxin B from the variant *Clostridium difficile* serotype F strain 1470. *Toxins (Basel)* **2013**, *5*, 106–119. [CrossRef] [PubMed]

15. Geny, B.; Grassart, A.; Manich, M.; Chicanne, G.; Payrastre, B.; Sauvonnet, N.; Popoff, M.R. Rac1 inactivation by lethal toxin from *Clostridium sordellii* modifies focal adhesions upstream of actin depolymerization. *Cell. Microbiol.* **2010**, *12*, 217–232. [CrossRef] [PubMed]

16. Genth, H.; Pauillac, S.; Schelle, I.; Bouvet, P.; Bouchier, C.; Varela-Chavez, C.; Just, I.; Popoff, M.R. Haemorrhagic toxin and lethal toxin from *Clostridium sordellii* strain vpi9048: Molecular characterization and comparative analysis of substrate specificity of the large clostridial glucosylating toxins. *Cell. Microbiol.* **2014**, *16*, 1706–1721. [CrossRef] [PubMed]

17. Lica, M.; Schulz, F.; Schelle, I.; May, M.; Just, I.; Genth, H. Difference in the biological effects of *Clostridium difficile* toxin B in proliferating and non-proliferating cells. *Naunyn Schmiedebergs Arch. Pharmacol.* **2011**, *383*, 275–283. [CrossRef] [PubMed]

18. Wohlan, K.; Goy, S.; Olling, A.; Srivaratharajan, S.; Tatge, H.; Genth, H.; Gerhard, R. Pyknotic cell death induced by *Clostridium difficile* TcdB: Chromatin condensation and nuclear blister are induced independently of the glucosyltransferase activity. *Cell. Microbiol.* **2014**, *16*, 1678–1692. [CrossRef] [PubMed]

19. D'Auria, K.M.; Donato, G.M.; Gray, M.C.; Kolling, G.L.; Warren, C.A.; Cave, L.M.; Solga, M.D.; Lannigan, J.A.; Papin, J.A.; Hewlett, E.L. Systems analysis of the transcriptional response of human ileocecal epithelial cells to *Clostridium difficile* toxins and effects on cell cycle control. *BMC. Syst. Biol.* **2012**, *6*, 2.

20. Arechederra, M.; Priego, N.; Vazquez-Carballo, A.; Sequera, C.; Gutierrez-Uzquiza, A.; Cerezo-Guisado, M.I.; Ortiz-Rivero, S.; Roncero, C.; Cuenda, A.; Guerrero, C.; Porras, A. p38 MAPK down-regulates fibulin 3 expression through methylation of gene regulatory sequences: Role in migration and invasion. *J. Biol. Chem.* **2015**, *290*, 4383–4397. [CrossRef] [PubMed]

21. Cerezo-Guisado, M.I.; del Reino, P.; Remy, G.; Kuma, Y.; Arthur, J.S.; Gallego-Ortega, D.; Cuenda, A. Evidence of p38gamma and p38delta involvement in cell transformation processes. *Carcinogenesis* **2011**, *32*, 1093–1099. [CrossRef] [PubMed]

22. Geny, B.; Popoff, M.R. Activation of a c-Jun-NH2-terminal kinase pathway by the lethal toxin from *Clostridium sordellii*, TcsL-82, occurs independently of the toxin intrinsic enzymatic activity and facilitates small GTPase glucosylation. *Cell. Microbiol.* **2009**, *11*, 1102–1113. [CrossRef] [PubMed]

23. Huelsenbeck, J.; May, M.; Schulz, F.; Schelle, I.; Ronkina, N.; Hohenegger, M.; Fritz, G.; Just, I.; Gerhard, R.; Genth, H. Cytoprotective effect of the small GTPase RhoB expressed upon treatment of fibroblasts with the Ras-glucosylating *Clostridium sordellii* lethal toxin. *FEBS Lett.* **2012**, *586*, 3665–3673. [CrossRef] [PubMed]

24. Schulz, F.; Just, I.; Genth, H. Prevention of *Clostridium sordellii* lethal toxin-induced apoptotic cell death by tauroursodeoxycholic acid. *Biochemistry* **2009**, *48*, 9002–9010. [CrossRef] [PubMed]

25. Ronkina, N.; Menon, M.B.; Schwermann, J.; Arthur, J.S.; Legault, H.; Telliez, J.B.; Kayyali, U.S.; Nebreda, A.R.; Kotlyarov, A.; Gaestel, M. Stress induced gene expression: A direct role for MAPKAP kinases in transcriptional activation of immediate early genes. *Nucleic Acids Res.* **2011**, *39*, 2503–2518. [CrossRef] [PubMed]

26. Dolado, I.; Swat, A.; Ajenjo, N.; de, V.G.; Cuadrado, A.; Nebreda, A.R. p38alpha MAP kinase as a sensor of reactive oxygen species in tumorigenesis. *Cancer Cell.* **2007**, *11*, 191–205. [CrossRef] [PubMed]

27. Cuadrado, A.; Nebreda, A.R. Mechanisms and functions of p38 MAPK signalling. *Biochem. J.* **2010**, *429*, 403–417. [CrossRef] [PubMed]

28. Halabi-Cabezon, I.; Huelsenbeck, J.; May, M.; Ladwein, M.; Rottner, K.; Just, I.; Genth, H. Prevention of the cytopathic effect induced by *Clostridium difficile* Toxin B by active Rac1. *FEBS Lett.* **2008**, *582*, 3751–3756. [CrossRef] [PubMed]

29. Genth, H.; Just, I. Functional implications of lethal toxin-catalysed glucosylation of (H/K/N)Ras and Rac1 in *Clostridium sordellii*-associated disease. *Eur. J. Cell Biol.* **2011**, *90*, 959–965. [CrossRef] [PubMed]

30. Genth, H.; Huelsenbeck, J.; Hartmann, B.; Hofmann, F.; Just, I.; Gerhard, R. Cellular stability of Rho-GTPases glucosylated by *Clostridium difficile* toxin B. *FEBS Lett.* **2006**, *580*, 3565–3569. [CrossRef] [PubMed]

31. Brandes, V.; Schelle, I.; Brinkmann, S.; Schulz, F.; Schwarz, J.; Gerhard, R.; Genth, H. Protection from *Clostridium difficile* toxin B-catalysed Rac1/Cdc42 glucosylation by tauroursodeoxycholic acid-induced Rac1/Cdc42 phosphorylation. *Biol. Chem.* **2012**, *393*, 77–84. [CrossRef] [PubMed]

32. May, M.; Schelle, I.; Brakebusch, C.; Rottner, K.; Genth, H. Rac1-dependent recruitment of PAK2 to G phase centrosomes and their roles in the regulation of mitotic entry. *Cell Cycle* **2014**, *13*, 2210–2220. [CrossRef] [PubMed]

33. Boehm, C.; Gibert, M.; Geny, B.; Popoff, M.R.; Rodriguez, P. Modification of epithelial cell barrier permeability and intercellular junctions by *Clostridium sordellii* lethal toxins. *Cell. Microbiol.* **2006**, *8*, 1070–1085. [CrossRef] [PubMed]

34. Popoff, M.R. Bacterial factors exploit eukaryotic Rho GTPase signaling cascades to promote invasion and proliferation within their host. *Small GTPases* **2014**, *5*, e28209. [CrossRef] [PubMed]

35. Zak, J.; Schneider, S.W.; Eue, I.; Ludwig, T.; Oberleithner, H. High-resistance MDCK-C7 monolayers used for measuring invasive potency of tumour cells. *Pflugers Arch.* **2000**, *440*, 179–183. [CrossRef] [PubMed]

36. Jank, T.; Giesemann, T.; Aktories, K. *Clostridium difficile* glucosyltransferase toxin B-essential amino acids for substrate binding. *J. Biol. Chem.* **2007**, *282*, 35222–35231. [CrossRef] [PubMed]

37. Felberbaum-Corti, M.; Cavalli, V.; Gruenberg, J. Capture of the small GTPase Rab5 by GDI: Regulation by p38 MAP kinase. *Methods Enzymol.* **2005**, *403*, 367–381. [PubMed]

38. Mace, G.; Miaczynska, M.; Zerial, M.; Nebreda, A.R. Phosphorylation of EEA1 by p38 MAP kinase regulates mu opioid receptor endocytosis. *EMBO J.* **2005**, *24*, 3235–3246. [CrossRef] [PubMed]

39. Zwang, Y.; Yarden, Y. p38 MAP kinase mediates stress-induced internalization of EGFR: Implications for cancer chemotherapy. *EMBO J.* **2006**, *25*, 4195–4206. [CrossRef] [PubMed]

40. Zhou, Y.; Tanaka, T.; Sugiyama, N.; Yokoyama, S.; Kawasaki, Y.; Sakuma, T.; Ishihama, Y.; Saiki, I.; Sakurai, H. p38-Mediated phosphorylation of Eps15 endocytic adaptor protein. *FEBS Lett.* **2014**, *588*, 131–137. [CrossRef] [PubMed]

41. Menon, M.B.; Kotlyarov, A.; Gaestel, M. SB202190-induced cell type-specific vacuole formation and defective autophagy do not depend on p38 MAP kinase inhibition. *PLoS ONE* **2011**, *6*, e23054. [CrossRef] [PubMed]

42. Menon, M.B.; Dhamija, S.; Kotlyarov, A.; Gaestel, M. The problem of pyridinyl imidazole class inhibitors of MAPK14/p38alpha and MAPK11/p38beta in autophagy research. *Autophagy* **2015**, *11*, 1425–1427. [CrossRef] [PubMed]

43. Genth, H.; Selzer, J.; Busch, C.; Dumbach, J.; Hofmann, F.; Aktories, K.; Just, I. New method to generate enzymatically deficient *Clostridium difficile* toxin B as an antigen for immunization. *Infect. Immun.* **2000**, *68*, 1094–1101. [CrossRef] [PubMed]

44. Popoff, M.R. Purification and characterization of *Clostridium sordellii* lethal toxin and cross-reactivity with *Clostridium difficile* cytotoxin. *Infect. Immun.* **1987**, *55*, 35–43. [PubMed]

45. Ambrosino, C.; Mace, G.; Galban, S.; Fritsch, C.; Vintersten, K.; Black, E.; Gorospe, M.; Nebreda, A.R. Negative feedback regulation of MKK6 mRNA stability by p38alpha mitogen-activated protein kinase. *Mol. Cell. Biol.* **2003**, *23*, 370–381. [CrossRef] [PubMed]

toxins

MDPI

Article

Semicarbazone EGA Inhibits Uptake of Diphtheria Toxin into Human Cells and Protects Cells from Intoxication

Leonie Schnell [1], Ann-Katrin Mittler [1], Andrea Mattarei [2], Domenico Azarnia Tehran [3], Cesare Montecucco [3] and Holger Barth [1,*]

[1] Institute of Pharmacology and Toxicology, University of Ulm Medical Center, Albert-Einstein-Allee 11, 89081 Ulm, Germany; leonie.schnell@uni-ulm.de (L.S.); ann-katrin.mittler@uni-ulm.de (A.-K.M.)
[2] Department of Chemical Sciences, University of Padova, 35121 Padova, Italy; andrea.mattarei@unipd.it
[3] Department of Biomedical Sciences, University of Padova, 35131 Padova, Italy; doazte@gmail.com (D.A.T.); cesare.montecucco@gmail.com (C.M.)
* Correspondence: holger.barth@uni-ulm.de; Tel.: +49-731-500-65503; Fax: +49-731-500-65502

Academic Editor: Jun Sakurai
Received: 6 May 2016; Accepted: 7 July 2016; Published: 15 July 2016

Abstract: Diphtheria toxin is a single-chain protein toxin that invades human cells by receptor-mediated endocytosis. In acidic endosomes, its translocation domain inserts into endosomal membranes and facilitates the transport of the catalytic domain (DTA) from endosomal lumen into the host cell cytosol. Here, DTA ADP-ribosylates elongation factor 2 inhibits protein synthesis and leads to cell death. The compound 4-bromobenzaldehyde *N*-(2,6-dimethylphenyl)semicarbazone (EGA) has been previously shown to protect cells from various bacterial protein toxins which deliver their enzymatic subunits from acidic endosomes to the cytosol, including *Bacillus anthracis* lethal toxin and the binary clostridial actin ADP-ribosylating toxins C2, iota and *Clostridium difficile* binary toxin (CDT). Here, we demonstrate that EGA also protects human cells from diphtheria toxin by inhibiting the pH-dependent translocation of DTA across cell membranes. The results suggest that EGA might serve for treatment and/or prevention of the severe disease diphtheria.

Keywords: diphtheria; diphtheria toxin; cellular uptake; membrane transport; EGA

1. Introduction

Corynebacterium diphtheriae produces the single-chain diphtheria toxin (DT, 58 kDa), which is the causative agent of diphtheria [1]. DT is efficiently taken up into human cells and its catalytic domain (DTA, 21 kDa) acts as an extremely potent enzyme in the cytosol. DTA covalently transfers ADP-ribose from cellular NAD^+ onto a modified histidine residue (diphthamide) of the elongation factor 2 (EF-2) thereby inhibiting protein synthesis and causing cell death [2,3], which can be monitored in terms of cell-rounding using HeLa cells [4,5]. DTA is located in the N-terminal domain of DT [6] while the C-terminal part (DTB, 37 kDa) mediates binding of the toxin to susceptible cells and the subsequent transport of DTA into the cytosol. DTB contains a receptor-binding (B) domain, which binds to the heparin-binding epidermal growth factor-like growth factor precursor (HB-EGF) [7,8] and a translocation (T) domain [9], which inserts into the membranes of acidified endosomes [10,11] allowing the membrane translocation of DTA from the endosomal lumen into the cytosol [12–18]. This process is prevented by bafilomycin A1, an inhibitor of endosomal acidification [19] and can be experimentally mimicked on the surface of cultured cells by exposure of cell-bound DT to an acidic pulse [20]. This triggers the insertion of DTB directly into the plasma membrane and the translocation of DTA into the cytosol where it modifies its substrate [21–23].

Translocation of DTA across endosomal membranes is facilitated by host cell factors including the chaperone heat shock protein (Hsp) 90 [24,25] and thioredoxin reductase [5,24,26]. DTA is separated from DTB by cleavage prior or during DT uptake [27] but these two subunits remain linked via an interchain disulfide between Cys-186 of DTA and Cys-201 of DTB [28]. The integrity of the interchain disulfide bond is essential during toxin uptake into endosomes, as well as DTA translocation across the membranes [27,29] but its reduction is necessary for the subsequent release of DTA on the cytosolic side [23] and this process is the rate-limiting step during DT uptake [30]. Reduction of the disulfide bond likely happens after membrane insertion of the T-domain [30] during or after DTA translocation to the cytosol [31]. Thioredoxin 1 reduces this disulfide under acidic conditions in vitro [32] and we recently demonstrated that pharmacological inhibition of thioredoxin reductase prevents DTA transport across cell membranes and protects cells from intoxication [5], implicating that this enzyme is crucial for the reduction of the disulfide bond and the subsequent release of DTA in the cell cytosol of living cells.

The compound 4-bromobenzaldehyde *N*-(2,6-dimethylphenyl)semicarbazone (EGA) was previously identified as a potent inhibitor that protects cells from toxins and viruses which enter the cytosol of cells via acidified vesicles [33,34], such as the *Bacillus anthracis* lethal toxin and DT [34], as well as the binary actin ADP-ribosylating toxins C2 from *Clostridium* (*C.*) *botulinum*, iota from *C. perfringens* and CDT from *C. difficile* [35]. EGA also protects neuronal cells from *C. botulinum* neurotoxins [36] and it was suggested that this compound might modulate intracellular toxin trafficking [34–36]. Prompted by these findings, we analyzed the effect of EGA on the intoxication of HeLa cells with DT in more detail. Here, we demonstrate that EGA significantly delays intoxication of cells with DT in a time- and concentration-dependent manner and analyzed the underlying molecular mechanism.

2. Results and Discussion

EGA protects HeLa cells from intoxication with DT. In a first set of experiments the possible inhibitory effect of EGA on the intoxication of HeLa cells by DT was investigated. To this end, cells were pre-incubated for 1 h with increasing concentrations of EGA and then challenged with DT. After different incubation periods, the number of round cells was determined because this is an established, highly specific and sensitive endpoint to monitor the intoxication process [5]. As shown in Figure 1, EGA significantly delayed the DT-induced cell-rounding in a time- and concentration-dependent manner, indicating that EGA interferes with the mode of action of DT in these cells. EGA delayed intoxication with DT even when cells were not grown to confluence and therefore more susceptible to DT. Importantly, EGA alone had no effects on the cells under such conditions (Figure 1A). Adverse effects on the cells were observed at concentrations of 500 μM EGA and above (not shown). The usage of 100 μM instead of 50 μM was still tolerated by the cells but did not result in a relevant improvement of the inhibitory effect (Figure 1B). Therefore, for the following investigation of the underlying molecular mechanism of the EGA-mediated protection of cells from DT, an EGA concentration of 50 μM was used. A prolonged pre-incubation period with EGA (up to 6 h) did not significantly enhance the inhibitory effect of EGA towards DT compared to an 1 h pre-incubation period. Moreover, EGA showed some protective effect towards DT, even when applied 15 min after DT (not shown).

EGA did not completely inhibit intoxication with DT but delayed it, which was expected from our earlier results with EGA and binary clostridial toxins [35] and with DT and other pharmacological inhibitors of toxin uptake [5]. The reason is most likely related to the extreme potency of AB-toxins in cells, as few molecules of their catalytic moieties in the host cell cytosol is usually sufficient to exhibit the full cytotoxic effects [37]. A pharmacological inhibition might therefore not be able to completely prevent the uptake of some DTA molecules into the cytosol over longer incubation periods.

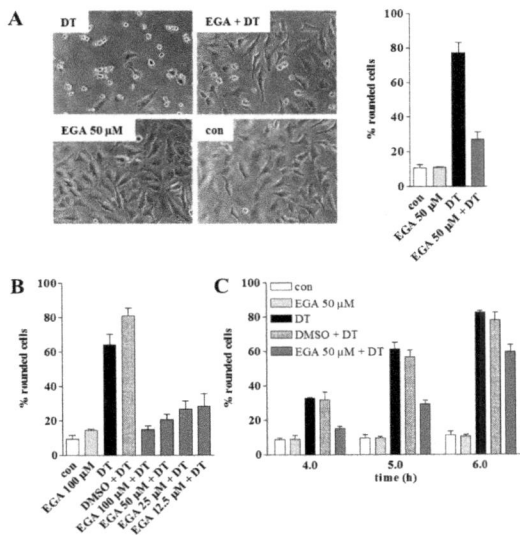

Figure 1. EGA protects HeLa cells from intoxication with DT. Cells were pre-incubated for 1 h at 37 °C with EGA or the solvent DMSO for control. Thereafter, DT (0.86 nM) was added and cells were further incubated at 37 °C. For control, cells were left untreated (con) or incubated with DT in the absence of EGA (DT). At the indicated time points, phase contrast pictures were taken and for quantitative analysis, the percentage of rounded cells was determined. Values are gives as mean ± SD (*n* = 3). (**A**) Protective effect of 50 μM EGA on the intoxication of cells with DT. Representative pictures after 4 h are shown; (**B**) Concentration-dependent inhibition of DT-intoxication after 4 h by EGA, determined in a separate experiment. The maximum dosage of EGA (100 μM) was tested on cells without DT to exclude any morphological alteration of the cells in response to EGA alone. In addition, a DMSO solvent control of the maximum dosage was performed; (**C**) Time-dependent inhibitory effect mediated by 50 μM EGA, determined in a separate experiment with confluent cells.

EGA inhibits the membrane transport of DTA in cells. The effect of EGA on the membrane transport of DTA was investigated by using a well-established assay where the endosomes acidic environment is mimicked on the surface of cultured HeLa cells. DT was allowed to bind to the receptor at 4 °C where no endocytosis occurs and the normal toxin uptake via acidic endosomes was inhibited by bafilomycin A1 treatment. When such cells are exposed to warm acidic medium, DTB inserts into the plasma membrane and mediates the translocation of DTA into the cytosol where it ADP-ribosylates EF-2 leading to cell-rounding, as shown in Figure 2. No cell-rounding was observed under neutral conditions, by incubation of cells in the acidic medium in the absence of DT, or in the presence of EGA without DT (Figure 2), indicating that cell-rounding is indeed specifically induced by the action of DTA, as described earlier [5,20]. Importantly, the DT-induced cell-rounding was significantly reduced when cells were pre-treated with EGA (Figure 2). Given that this inhibitor did not inhibit the in vitro enzymatic activity of DTA (Figure 3A) nor the binding of DT to HeLa cells (Figure 3B) and it leads to inhibition of intoxication also when applied after receptor-binding of nicked DT (i.e., DT that was proteolytically activated in vitro prior to its application to the cells) (Figure 3C), our results strongly suggest that EGA interferes with the toxin pH-dependent transport. In addition, the results obtained by performing the translocation assay with nicked DT also exclude an effect of EGA on the proteolytic activation of DT during cellular uptake on the cell surface and/or in endosomal vesicles. To demonstrate that the observed inhibitory effect of EGA was mediated by the molecule itself, in each experiment a vehicle control (DMSO) was performed in parallel to EGA treatment.

In contrast to EGA, DMSO did not inhibit the intoxication of cells with DT (see Figure 1B,C) or the translocation of cell-bound DT (Figure 2).

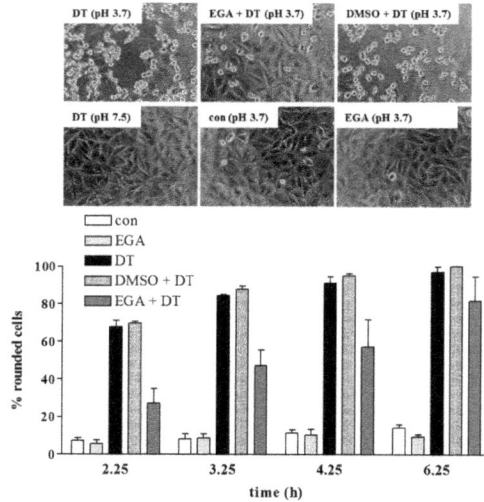

Figure 2. EGA inhibits the pH-dependent transport of the enzyme component DTA of DT across the cytoplasmic membranes in living cells. HeLa cells were pre-incubated for 1 h at 37 °C in serum-free medium with 100 nM bafilomycin (Baf) A1. In addition, some cells were also treated with 50 µM EGA or the corresponding solvent DMSO for control. Subsequently, cells were kept on ice for 10 min. Thereafter, nicked DT (13 nM) was added and cells were incubated for 40 min at 4 °C to enable toxin-binding to the receptors on the cell surface. Cells were further incubated for 15 min at 37 °C and pH 3.7 to trigger the pH-driven membrane transport of DTA. For control (con), cells were incubated at pH 7.5 (not shown) or at pH 3.5 (shown). During this step, there was no EGA present in the medium. Thereupon, all cells were further incubated at 37 °C in neutral medium (pH 7.5) containing serum, Baf A1 ± EGA. At the indicated time points, phase contrast pictures were taken and for quantitative analysis, the percentage of rounded cells was determined. Representative pictures are shown after 3.25 h. Values are gives as mean ± SD (n = 3).

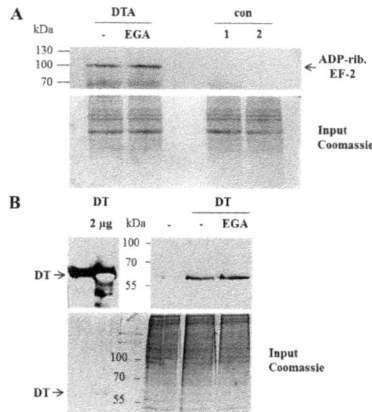

Figure 3. *Cont.*

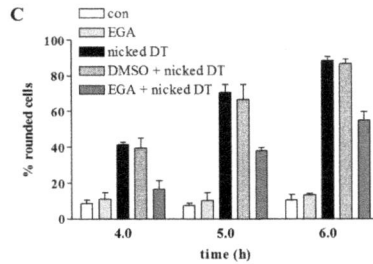

Figure 3. Treatment with EGA has no effect on the in vitro enzymatic activity of DTA nor on the receptor-binding of DT to cells. (**A**) DTA (100 ng) was pre-incubated with 50 μM EGA for 10 min at 37 °C. For control (con), a sample of DTA was left untreated. Thereupon, HeLa lysate protein (20 μg) as well as biotin-labelled NAD^+ were added followed by an incubation of the samples for 10 min at 37 °C. Thereafter, samples were subjected to SDS-PAGE, blotted and biotinylated (i.e., ADP-ribosylated, indicated as ADP-rib.) EF-2 was detected (upper panel). A sample of HeLa lysate with biotin-labelled NAD^+ (1) and one of only HeLa lysate (2) were additionally analyzed. As control of equal protein loading, the SDS-gel was stained with Coomassie after the blotting process (lower panel). Noteworthy, there was no obvious difference in the amount of ADP-ribosylated EF-2 when lysate from EGA-treated cells was used (not shown), indicating that EGA-treatment does not interfere with the generation of the diphthamide in EF-2; (**B**) HeLa cells were pre-treated with EGA (50 μM) for 1 h at 37 °C in serum-free medium or left untreated for control. Thereupon, cells were kept on ice for 10 min. Then, DT (34.4 nM) was added followed by further incubation for 1 h 10 min at 4 °C to enable binding of DT to the cell surface. Thereafter, three washing steps with ice-cold PBS were performed to remove unbound toxin and cells were scraped off in 2.5-fold SDS sample buffer containing 10% DTT. Proteins were separated by SDS-PAGE, blotted and bound DT was detected using a specific primary and a peroxidase-coupled secondary antibody and the ECL system (upper panel). For loading control, two dilutions of DT were analyzed on the left side of the same blot. Comparable amounts of total protein were confirmed by Coomassie staining of the proteins in the SDS-gel after the blotting process (lower panel); (**C**) Cells were kept on ice for 10 min before addition of nicked DT (3.4 nM) and further incubation at 4 °C for 30 min. Thereafter, the medium was removed and warm, serum-free medium containing EGA (50 μM) or DMSO was added followed by further incubation at 37 °C. For control (con), cells were left untreated or treated only with EGA without toxin. At the indicated time points, phase contrast pictures were taken and for quantitative analysis, the percentage of rounded cells was determined. Values are gives as mean $\pm$ SD (n = 3).

3. Conclusions

In conclusion, our results demonstrate that EGA protects cultured human cells from intoxication with DT because it interferes with the uptake of DTA into the host cell cytosol. Moreover, our findings suggest that EGA inhibits the pH-dependent transport of DTA across membranes into the cytosol. This step was also inhibited by EGA in the case of clostridial binary toxins, which also deliver their A subunit from acidic endosomes to the cytosol [35]. All toxins which were inhibited by EGA, exploit acidic endosomal vesicles for cellular trafficking, suggesting a common inhibitory mode of action for EGA in cells. The binary anthrax toxins and the binary clostridial actin ADP-ribosylating toxins share an overall comparable mechanism to deliver their enzymatically active components through pores formed by their separate heptameric transport components across endosomal membranes into the cytosol. Thus, a common mechanism by which EGA inhibits the membrane transport of these binary toxins seems plausible. However, EGA inhibits the membrane transport of DT, as shown in this present study, but it did not inhibit the translocation of botulinum neurotoxins across the plasma membrane [36], suggesting that the protective effect of EGA towards both single-chain toxins might not be mediated via the same molecular mechanism. Thus far, the molecular mechanisms by which

EGA interferes with the membrane transport of DT are not known and it can be speculated that EGA might block the transmembrane pore formed by the translocation domain of DT or might inhibit host cell factors which are involved in translocation of DTA across endosomal membranes.

However, although the precise mode of action how EGA inhibits the transport of the enzymatically active moieties of bacterial toxins into the cytosol requires further elucidation, this compound inhibits a series of the most powerful and medically most relevant bacterial toxins including DT, botulinum neurotoxins and anthrax toxin, and thus might have very attractive clinical applications. In previous studies [36] the treatment of mice with EGA by multiple intraperitoneally injections were well tolerated. It should be taken into account that higher concentrations of DT might be used in the experimental approach compared to the situation in patients. Moreover, EGA could represent a lead compound and chemical modifications may result in even more efficient compounds for future clinical applications.

4. Experimental Section

4.1. Materials and Reagents

Cell culture media (MEM) and fetal calf serum (FCS) were obtained from Invitrogen (Karlsruhe, Germany) and cell culture materials from TPP (Trasadingen, Switzerland). The protein molecular weight marker Page Ruler prestained Protein ladder® was purchased from Thermo Fisher Scientific Inc. (Waltham, MA, USA). Complete® protease inhibitor was supplied by Roche (Mannheim, Germany). Biotinylated NAD⁺ was from R&D Systems GmbH (Wiesbaden-Nordenstadt, Germany). Bafilomycin (Baf) A1 was purchased from Calbiochem (Bad Soden, Germany), 2-(4-bromobenzylidene)-*N*-(2,6-dimethylphenyl)hydrazinecarboxamide (EGA) was synthesized as described [36]. The quality of EGA was tested by HPLC-MS analysis. The purity is >95% [36].

Streptavidin-peroxidase was obtained from Roche (Mannheim, Germany), the enhanced chemiluminescence (ECL) system from Millipore (Schwalbach, Germany) and the nitrocellulose blotting membrane from Whatman® (Dassel, Germany). Diphtheria toxin antibody was obtained from GeneTex Inc. (Irvine, CA, USA) and goat anti-mouse IgG-HRP secondary antibody from Santa-Cruz Biotechnology Inc. (Heidelberg, Germany). Diphtheria toxin (DT) was supplied by Calbiochem (Bad Soden, Germany). DT was proteolytically activated in vitro with trypsin to yield nicked DT as described in Section 4.4. and DTA was expressed and purified as described earlier [5].

4.2. Cell Culture and Cytotoxicity Assays

HeLa cells were cultivated at 37 °C and 5% CO_2 in MEM containing 10% heat-inactivated FCS, 1 mM sodium-pyruvate, 2 mM L-glutamine, 0.1 mM non-essential amino acids and 1% penicillin-streptomycin. Cells were trypsinized and reseeded for at most 30 times.

For cytotoxicity experiments, cells were grown in 96-well plates and pre-incubated with EGA or the vehicle DMSO in serum-free medium for 1 h at 37 °C. For control, cells were incubated without inhibitor. Thereafter, DT (0.86 nM) was added and cells were further incubated with toxin plus inhibitor at 37 °C. For analysis of the cellular effect of EGA when added after DT-binding to cells, cells grown in 96-well plates were kept on ice for 10 min in serum-free medium before addition of nicked DT (3.4 nM). After incubation for 30 min at 4 °C, medium was removed and warm, serum-free medium containing EGA (50 μM) or the vehicle DMSO was added followed by further incubation at 37 °C. After the indicated incubation periods, a Zeiss Axiovert 40 CFL microscope (Oberkochen, Germany) with a Jenoptik progress C10 CCD camera (Jena, Germany) was used to visualize the cells to analyze the DT-induced morphological changes. The characteristic rounding up of the cells was taken as specific endpoint to monitor the intoxication process. For quantitative analysis, cells were counted per picture (ImageJ software; NIH, Bethesda, MD, USA) and the amount of rounded cells was determined in percent.

4.3. SDS-PAGE and Western Blotting

For immunoblot analysis, according to the method of Laemmli [38], equal amounts of protein were subjected to SDS-PAGE. Afterwards, the proteins were transferred to a nitrocellulose membrane which was then blocked for 1 h at RT with 5% dry milk powder in PBS containing 0.1% Tween-20 (PBS-T) or alternatively overnight at 4 °C. For detection of the biotin-labelled EF-2 or DT, the samples were either probed with streptavidin-peroxidase or a specific antibody against DTA, respectively, for 1 h followed by washing steps with PBS-T. Subsequently, in case of DT, the membrane was probed with anti-mouse secondary antibody coupled to horseradish peroxidase for 1 h followed by further washing steps with PBS-T. Thereafter, the proteins were visualized using the ECL system according to the manufacturer's instructions. Ponceau S staining of the membrane and Coomassie staining of the gel were used to confirm equal amounts of protein.

4.4. Proteolytic Activation (Nicking) of DT

To activate DT, the toxin was treated with trypsin (3 μg/mL) for 3 h at 37 °C and then kept on ice. Following this, trypsin was neutralized by incubation with trypsin inhibitor (30 μg/mL) for 80 min at 4 °C and the concentration of nicked DT was determined using SDS-PAGE.

4.5. ADP-Ribosylation of EF-2 by DTA in a Cell-Free System

DTA (100 ng) was pre-incubated for 10 min at 37 °C with EGA (50 μM) or left untreated for control followed by the addition of HeLa lysate protein (20 μg) and biotin-labelled NAD$^+$ (10 μM). After further incubation of the samples for 10 min at 37 °C, the proteins were subjected to SDS-PAGE, blotted onto a nitrocellulose membrane. Finally, biotin-labelled (i.e., ADP-ribosylated) EF-2 was detected using Western blotting.

4.6. Binding of DT to Its Cell Surface Receptor

HeLa cells were pre-incubated in serum-free medium with EGA (50 μM) for 1 h at 37 °C. For control, cells were left untreated. Following this, cells were kept at 4 °C for 10 min. Then, DT (34.4 nM) was added and cells were further incubated at 4 °C for 1 h and 10 min to enable binding of DT to its specific cell surface receptor. Three washing steps with ice-cold PBS were used to remove unbound toxin and cells were subsequently scraped in 2.5-fold concentrated sample buffer containing 10% DTT [38]. SDS-PAGE was used for protein separation and after blotting, bound DT was detected using a specific antibody against DT and peroxidase-coupled secondary anti-mouse antibody.

4.7. DTA Translocation Assay across the Cytoplasmic Membrane of Living Cells

The pH-dependent translocation of DT across endosomal membranes was experimentally mimicked on the cytoplasmic membranes of intact cells and performed as described earlier [5]. In brief, HeLa cells were pre-incubated in serum-free medium with Baf A1 (100 nM) plus EGA (50 μM) or the vehicle DMSO for 1 h at 37 °C. Then, the cells were kept on ice for 10 min followed by an incubation with nicked DT (13 nM) for 40 min at 4 °C to allow binding of the toxin to the cell surface. Thereafter, to trigger pH-driven toxin-translocation across the surface membrane, cells were exposed to an acidic pulse (pH 3.7) for 15 min at 37 °C. Additional wells with toxin-treated cells were treated with neutral medium (pH 7.5) for control. Subsequently, all cells were further incubated at 37 °C in neutral medium containing FCS, Baf A1 (100 nM) ± EGA (50 μM). After the given incubation periods, DT-induced cell-rounding was monitored using photography. From these, the amount of rounded cells was determined in per cent of total cell count per picture.

4.8. Reproducibility of the Experiments

All experiments were independently performed at least two times and in the figures, results from representative experiments are shown. Quantification was performed by calculating the values

($n = 3$; $n = 3$ refers to three fields of view in one experiment and several independent experiments were performed, all with comparable results) as the means $\pm$ standard deviation (S.D.) using the Prism4 Software (Version 4.0, GraphPad Software Inc., La Jolla, CA, USA, 2003).

Acknowledgments: The work was financially supported by the Deutsche Forschungsgemeinschaft DFG (grant BA 2087/2-2). Leonie Schnell is a member of the International School in Molecular Medicine Ulm (IGradU) and thanks IGradU for the support.

Author Contributions: Conceived and designed the experiments: H.B., L.S.; Performed the experiments: L.S., A.-K.M.; Analyzed the data: H.B., L.S.; Wrote the paper: H.B., L.S.; Contributed essential materials, reagents, tools: A.M., D.A.T., C.M.

Conflicts of Interest: The authors declare no conflict of interest.

References

1. Murphy, J.R. Mechanism of diphtheria toxin catalytic domain delivery to the eukaryotic cell cytosol and the cellular factors that directly participate in the process. *Toxins* **2011**, *3*, 294–308. [CrossRef] [PubMed]
2. Pappenheimer, A.M. Diphtheria toxin. *Annu. Rev. Biochem.* **1977**, *46*, 69–94. [CrossRef] [PubMed]
3. Collier, R.J.; Cole, H.A. Diphtheria toxin subunit active in vitro. *Science* **1969**, *164*, 1179–1181. [CrossRef] [PubMed]
4. Blumenthal, B.; Hoffmann, C.; Aktories, K.; Backert, S.; Schmidt, G. The cytotoxic necrotizing factors from Yersinia pseudotuberculosis and from Escherichia coli bind to different cellular receptors but take the same route to the cytosol. *Infect. Immun.* **2007**, *75*, 3344–3353. [CrossRef] [PubMed]
5. Schnell, L.; Dmochewitz-Kück, L.; Feigl, P.; Montecucco, C.; Barth, H. Thioredoxin reductase inhibitor auranofin prevents membrane transport of diphtheria toxin into the cytosol and protects human cells from intoxication. *Toxicon* **2016**, *116*, 23–28. [CrossRef] [PubMed]
6. Collier, R.J.; Kandel, J. Structure and activity of diphtheria toxin. I. Thiol-dependent dissociation of a fraction of toxin into enzymatically active and inactive fragments. *J. Biol. Chem.* **1971**, *246*, 1496–1503. [PubMed]
7. Iwamoto, R.; Higashiyama, S.; Mitamura, T.; Taniguchi, N.; Klagsbrun, M.; Mekada, E. Heparin-binding EGF-like growth factor, which acts as the diphtheria toxin receptor, forms a complex with membrane protein DRAP27/CD9, which up-regulates functional receptors and diphtheria toxin sensitivity. *EMBO J.* **1994**, *13*, 2322–2330. [PubMed]
8. Naglich, J.G.; Metherall, J.E.; Russell, D.W.; Eidels, L. Expression cloning of a diphtheria toxin receptor: Identity with a heparin-binding EGF-like growth factor precursor. *Cell* **1992**, *69*, 1051–1061. [CrossRef]
9. Choe, S.; Bennett, M.J.; Fujii, G.; Curmi, P.M.; Kantardjieff, K.A.; Collier, R.J.; Eisenberg, D. The crystal structure of diphtheria toxin. *Nature* **1992**, *357*, 216–222. [CrossRef] [PubMed]
10. Johnson, V.G.; Nicholls, P.J.; Habig, W.H.; Youle, R.J. The role of proline 345 in diphtheria toxin translocation. *J. Biol. Chem.* **1993**, *268*, 3514–3519. [PubMed]
11. Perier, A.; Chassaing, A.; Raffestin, S.; Pichard, S.; Masella, M.; Ménez, A.; Forge, V.; Chenal, A.; Gillet, D. Concerted protonation of key histidines triggers membrane interaction of the diphtheria toxin T domain. *J. Biol. Chem.* **2007**, *282*, 24239–24245. [CrossRef] [PubMed]
12. Drazin, R.; Kandel, J.; Collier, R.J. Structure and activity of DT II. Attack by trypsin at a specific site within the intact molecule. *J. Biol. Chem.* **1971**, *246*, 1504–1510. [PubMed]
13. Papini, E.; Sandoná, D.; Rappuoli, R.; Montecucco, C. On the membrane translocation of diphtheria toxin: At low pH the toxin induces ion channels on cells. *EMBO J.* **1988**, *7*, 3353–3359. [PubMed]
14. Lemichez, E.; Bomsel, M.; Devilliers, G.; vanderSpek, J.; Murphy, J.R.; Lukianov, E.V.; Olsnes, S.; Boquet, P. Membrane translocation of diphtheria toxin fragment A exploits early to late endosome trafficking machinery. *Mol. Microbiol.* **1997**, *23*, 445–457. [CrossRef] [PubMed]
15. Wiedlocha, A.; Madshus, I.H.; Mach, H.; Middaugh, C.R.; Olsnes, S. Tight folding of acidic fibroblast growth factor prevents its translocation to the cytosol with diphtheria toxin as vector. *EMBO J.* **1992**, *11*, 4835–4842. [PubMed]
16. Falnes, P.O.; Choe, S.; Madshus, I.H.; Wilson, B.A.; Olsnes, S. Inhibition of membrane translocation of diphtheria toxin A-fragment by internal disulfide bridges. *J. Biol. Chem.* **1994**, *269*, 8402–8407. [PubMed]

17. Stenmark, H.; Olsnes, S.; Madshus, I.H. Elimination of the disulphide bridge in fragment B of diphtheria toxin: Effect on membrane insertion, channel formation, and ATP binding. *Mol. Microbiol.* **1991**, *5*, 595–606. [CrossRef] [PubMed]

18. Leka, O.; Vallese, F.; Pirazzini, M.; Berto, P.; Montecucco, C.; Zanotti, G. Diphtheria toxin conformational switching at acidic pH. *FEBS J.* **2014**, *281*, 2115–2122. [CrossRef] [PubMed]

19. Umata, T.; Moriyama, Y.; Futai, M.; Mekada, E. The cytotoxic action of diphtheria toxin and its degradation in intact Vero cells are inhibited by bafilomycin A1, a specific inhibitor of vacuolar-type H(+)-ATPase. *J. Biol. Chem.* **1990**, *265*, 21940–21945. [PubMed]

20. Sandvig, K.; Olsnes, S. Rapid entry of nicked diphtheria toxin into cells at low pH. Characterization of the entry process and effects of low pH on the toxin molecule. *J. Biol. Chem.* **1981**, *256*, 9068–9076. [PubMed]

21. Sandvig, K.; Olsnes, S. Diphtheria toxin-induced channels in Vero cells selective for monovalent cations. *J. Biol. Chem.* **1988**, *263*, 12352–12359. [PubMed]

22. Moskaug, J.O.; Stenmark, H.; Olsnes, S. Insertion of diphtheria toxin B-fragment into the plasma membrane at low pH. Characterization and topology of inserted regions. *J. Biol. Chem.* **1991**, *266*, 2652–2659. [PubMed]

23. Falnes, P.O.; Madshus, I.H.; Sandvig, K.; Olsnes, S. Replacement of negative by positive charges in the presumed membrane-inserted part of diphtheria toxin B fragment. Effect on membrane translocation and on formation of cation channels. *J. Biol. Chem.* **1992**, *267*, 12284–12290. [PubMed]

24. Ratts, R.; Zeng, H.; Berg, E.A.; Blue, C.; McComb, M.E.; Costello, C.E.; vanderSpek, J.C.; Murphy, R. The cytosolic entry of diphtheria toxin catalytic domain requires a host cell cytosolic translocation factor complex. *J. Cell Biol.* **2003**, *160*, 1139–1150. [CrossRef] [PubMed]

25. Dmochewitz, L.; Lillich, M.; Kaiser, E.; Jennings, L.D.; Lang, A.E.; Buchner, J.; Fischer, G.; Aktories, K.; Collier, R.J.; Barth, H. Role of CypA and Hsp90 in membrane translocation mediated by anthrax protective antigen. *Cell. Microbiol.* **2011**, *13*, 359–373. [CrossRef] [PubMed]

26. Mustacich, D.; Powis, G. Thioredoxin reductase. *Biochem. J.* **2000**, *346*, 1–8. [CrossRef] [PubMed]

27. Tsuneoka, M.; Nakayama, K.; Hatsuzawa, K.; Komada, M.; Kitamura, N.; Mekada, E. Evidence for the involvement of furin in cleavage and activation of diphtheria toxin. *J. Biol. Chem.* **1993**, *268*, 26461–26465. [PubMed]

28. Gill, D.M.; Pappenheimer, A.M. Structure activity relationships in diphtheria toxin. *J. Biol. Chem.* **1971**, *246*, 1485–1491. [PubMed]

29. Ariansen, S.; Afanasiev, B.N.; Moskaug, J.O.; Stenmark, H.; Madhaus, I.H.; Olsnes, S. Membrane translocation of diphtheria toxin A-fragment: Role of carboxy-terminal region. *Biochemistry* **1993**, *32*, 83–90. [CrossRef] [PubMed]

30. Papini, E.; Rappuoli, R.; Murgia, M.; Montecucco, C. Cell penetration of diphtheria toxin. Reduction of the interchain disulfide bridge is the rate-limiting step of translocation in the cytosol. *J. Biol. Chem.* **1993**, *268*, 1567–1574. [PubMed]

31. Madshus, I.H.; Wiedlocha, A.; Sandvig, K. Intermediates in translocation of diphtheria toxin across the plasma membrane. *J. Biol. Chem.* **1994**, *269*, 4648–4652. [PubMed]

32. Moskaug, J.O.; Sandvig, K.; Olsnes, S. Cell-mediated reduction of the interfragment disulfide in nicked DT. A new system to study toxin entry at low pH. *J. Biol. Chem.* **1987**, *262*, 10339–10345. [PubMed]

33. Young, J.A.; Collier, R.J. Anthrax toxin: Receptor binding, internalization, pore formation, and translocation. *Annu. Rev. Biochem.* **2007**, *76*, 243–265. [CrossRef] [PubMed]

34. Gillespie, E.J.; Ho, C.L.; Balaji, K.; Clemens, D.L.; Deng, G.; Wang, Y.E.; Elsaesser, H.J.; Tamilselvam, B.; Gargi, A.; Dixon, S.D.; et al. Selective inhibitor of endosomal trafficking pathways exploited by multiple toxins and viruses. *Proc. Natl. Acad. Sci. USA* **2013**, *110*, E4904–E4912. [CrossRef] [PubMed]

35. Schnell, L.; Mittler, A.K.; Sadi, M.; Popoff, M.R.; Schwan, C.; Aktories, K.; Mattarei, A.; Tehran, D.A.; Montecucco, C.; Barth, H. EGA protects mammalian cells from *Clostridium difficile* CDT, *Clostridium perfringens* iota toxin, and *Clostridium botulinum* C2 toxin. *Toxins* **2016**, *8*, 101. [CrossRef] [PubMed]

36. Tehran, D.A.; Zanetti, G.; Leka, O.; Lista, F.; Fillo, S.; Binz, T.; Shone, C.C.; Rossetto, O.; Montecucco, C.; Paradisi, C.; et al. A Novel inhibitor prevents the peripheral neuroparalysis of botulinum neurotoxins. *Sci. Rep.* **2015**, *5*, 17513. [CrossRef] [PubMed]

37. Yamaizumi, M.; Mekada, E.; Uchida, T.; Okada, Y. One molecule of diphtheria toxin fragment A introduced into a cell can kill the cell. *Cell* **1978**, *15*, 245–250. [CrossRef]

38. Laemmli, U.K. Cleavage of structural proteins during the assembly of the head of bacteriophage T4. *Nature* **1970**, *227*, 680–685. [CrossRef] [PubMed]

Article

Impact of Dendrimer Terminal Group Chemistry on Blockage of the Anthrax Toxin Channel: A Single Molecule Study

Goli Yamini, Nnanya Kalu and Ekaterina M. Nestorovich *

Department of Biology, The Catholic University of America, Washington DC, WA 20064, USA;
00yamini@cua.edu (G.Y.); 56kalu@cua.edu (N.K.)
* Correspondence: nestorovich@cua.edu; Tel.: +1-202-319-6723

Academic Editor: Holger Barth
Received: 18 October 2016; Accepted: 7 November 2016; Published: 15 November 2016

Abstract: Nearly all the cationic molecules tested so far have been shown to reversibly block K^+ current through the cation-selective PA_{63} channels of anthrax toxin in a wide nM–mM range of effective concentrations. A significant increase in channel-blocking activity of the cationic compounds was achieved when multiple copies of positively charged ligands were covalently linked to multivalent scaffolds, such as cyclodextrins and dendrimers. Even though multivalent binding can be strong when the individual bonds are relatively weak, for drug discovery purposes we often strive to design multivalent compounds with high individual functional group affinity toward the respective binding site on a multivalent target. Keeping this requirement in mind, here we perform a single-channel/single-molecule study to investigate kinetic parameters of anthrax toxin PA_{63} channel blockage by second-generation (G2) poly(amido amine) (PAMAM) dendrimers functionalized with different surface ligands, including G2-NH$_2$, G2-OH, G2-succinamate, and G2-COONa. We found that the previously reported difference in IC_{50} values of the G2-OH/PA_{63} and G2-NH$_2$/PA_{63} binding was determined by both on- and off-rates of the reversible dendrimer/channel binding reaction. In 1 M KCl, we observed a decrease of about three folds in k_{on} and a decrease of only about ten times in t_{res} with G2-OH compared to G2-NH$_2$. At the same time for both blockers, k_{on} and t_{res} increased dramatically with transmembrane voltage increase. PAMAM dendrimers functionalized with negatively charged succinamate, but not carboxyl surface groups, still had some residual activity in inhibiting the anthrax toxin channels. At 100 mV, the on-rate of the G2-succinamate binding was comparable with that of G2-OH but showed weaker voltage dependence when compared to G2-OH and G2-NH$_2$. The residence time of G2-succinamate in the channel exhibited opposite voltage dependence compared to G2-OH and G2-NH$_2$, increasing with the *cis*-negative voltage increase. We also describe kinetics of the PA_{63} ion current modulation by two different types of the "imperfect" PAMAM dendrimers, the mixed-surface G2 75% OH 25% NH$_2$ dendrimer and G3-NH$_2$ dendron. At low voltages, both "imperfect" dendrimers show similar rate constants but significantly weaker voltage sensitivity when compared with the intact G2-NH$_2$ PAMAM dendrimer.

Keywords: multivalency; planar lipid bilayer technique; *Bacillus anthracis*; protective antigen; pore blockage

1. Introduction

Many bacterial exotoxins oligomerize during invasion to form ion-conductive channels or pores in the host cell or organelle membranes. This oligomeric centrosymmetric organization represents an ideal multivalent receptor target to explore a variety of multivalent channel-blocking ligands with a controlled number of preassembled functional groups (reviewed in [1]). In the past decade, two classes of multivalent compounds, cationic cyclodextrins (CDs) and dendrimers have been reported to directly block the channel-forming B components of the AB type anthrax, C2, iota, and CDT toxins [2–5], PA_{63}, C2IIa, Ib, and CDTb, respectively. The potency of these multivalent blockers compares well with the most effective blockers of the classical channels of electrophysiology (Table 3 in ref. [6]) and exceeds activities of the small-molecule cationic ligands [7–11]. A key advantage of cyclodextrins (reviewed in [12]) is their arrangement into rigid 6-, 7-, and 8-fold centrosymmetric structures with controlled number and position of potential attachment sites and the ability to form water-soluble "host-guest" inclusion complexes with poorly soluble small molecules and macromolecule fragments. The poly(amido amine) (PAMAM) dendrimers (reviewed in [13]) are repeatedly branched polymers with all bonds forming amidoamine branches emanating from a central alkyldiamine core, where each consecutive growth step represents a new dendrimer "generation" with an increased diameter and doubled number of reactive surface functional groups. PAMAM dendrimers form monodispersed, starburst-shaped polymers that are synthesized in generations with a growing but well-controlled number of attachment sites (Figure S1). While many details of the biophysical mechanisms of cyclodextrin interaction with the channel-forming components of anthrax, C2, and iota AB type toxins are known [14,15], the physical forces involved in the dendrimer/channel binding reaction require further analysis. In this small-scale study written for the special issue of Toxins (Basel) on "Novel Pharmacological Inhibitors for Bacterial Protein Toxins", we perform a single channel investigation of the PAMAM dendrimer/PA_{63} channel binding reaction focusing on two specific aspects of the dendrimer-induced channel blockage. Firstly, we perform a single-channel analysis of the kinetic parameters of the dendrimer/PA_{63} binding reaction using generation 2 amino-terminated (G2-NH_2), hydroxyl-terminated (G2-OH), succinamate-terminated (G2-SA), and carboxyl-terminated (G2-COONa) PAMAM dendrimers (Figure 1A and Figure S1). The amino-terminated PAMAM dendrimers of different generations were recently reported to effectively block the PA_{63} channel lumen in multichannel experiments under near-physiological conditions [5]. This effect was explained by the direct electrostatic interaction of the positively-charged terminal amino-groups on the PAMAM dendrimers with the negatively charged lumen of PA_{63}. However, second and third generation (G2 and G3) PAMAM dendrimers functionalized with surface hydroxyl groups were reported to have an avidity decrease of only 20 and 9 times compared to the G2 and G3 amino-terminated dendrimers, respectively. This difference is comparable with the variation observed between different PAMAM dendrimer generations. Moreover, even PAMAM dendrimers terminated with negatively charged carboxylic and succinamic surface groups showed some residual PA_{63} binding. When fine-tuned, some of these dendrimers may offer the advantage of being effective channel inhibitors with decreased cytotoxicity. In fact, while all dendrimers are less toxic than linear polymers [16,17], the cationic amino-terminated PAMAM dendrimers have been reported to display concentration- and generation-dependent cytotoxicity, and are therefore less biocompatible compared to their neutral and negatively charged analogues [18,19]. Secondly, we analyze the kinetic parameters of the blockage of a single PA_{63} channel by G2 75% OH 25% NH_2 PAMAM dendrimer and G3-NH_2 dendron (Figure 1B). It has been demonstrated that a favorable therapeutic window for dendrimers can also be achieved by either partial surface modification, aiming to lower amino group density, or by degradation of the dendrimers to "imperfect" dendrimers or fractured dendron-like branches [20]. The channel-blocking activity of these two types of the "imperfect" dendrimers was previously investigated on a multichannel level [5]. The mixed-surface G2 75% OH 25% NH_2 PAMAM dendrimer that, on average, had only four surface positive charges was about 17 times less active (IC_{50} = 122 nM vs. 7 nM) than the 16+ charged G2-NH_2 dendrimer, and its activity was comparable (IC_{50} = 122 nM vs. 128 nM) with the 4+ charged G0 dendrimer. The structurally incomplete 4+ charged G1 dendron was about 26 times more effective (IC_{50} = 4.9 nM) than the 4+ charged G0 dendrimer (IC_{50} = 128 nM). The 8+ charged G2 dendron and G1 dendrimer had similar inhibitory activity. The five different commercially available generation 2 dendrimers and one generation 3 dendron, investigated in this study, were chosen with the purpose of specifically focusing on blocker terminal group chemistry and dendrimer flexibility. The number of terminal amino, hydroxyl, succinamate, carboxyl, or mixed OH/NH$_2$ surface groups was fixed and equal to 16.

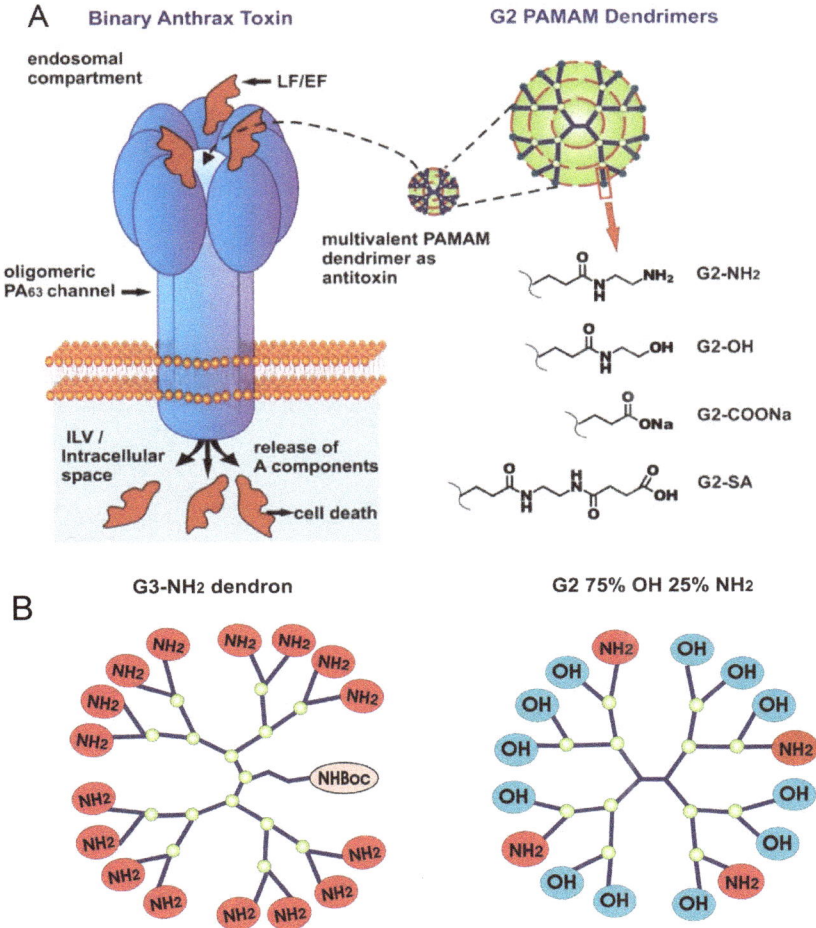

Figure 1. Inhibition of anthrax toxin by poly(amido amine) (PAMAM) dendrimers. (**A**) Oligomeric PA$_{63}$ channel, produced by *B. anthracis* (left), is responsible for translocation of lethal factor (LF) and edema factor (EF) into the host cell. The cartoon is a simplified illustration of the early/late endosomal stages of the LF and EF host cell transport. Second generation (G2) PAMAM dendrimers (right) tested in this study. All shown G2 dendrimers have 16 terminal groups that could be charged positively (G2-NH$_2$), negatively (G2-COONa, G2-SA) or are neutral (G2-OH); (**B**) Schematic representation of the mixed-surface G3-NH$_2$ dendron (left) and G2 75% OH 25% NH$_2$ dendrimer (right). Note that in contrast to all other dendrimers, G2 75% OH 25% NH$_2$ is not monodisperse and contains 75% of terminal OH groups and 25% of terminal NH$_2$ groups on average.

In this study, we investigated the oligomeric channel-forming B component of the anthrax toxin, protective antigen (PA), as a multivalent target for the multivalent dendrimer binding. Traditionally, PA, because of its main role in anthrax toxin uptake, has been one of the key targets for small molecule and multivalent antitoxin development [21]. The AB type anthrax toxin is composed of three individually nontoxic proteins. The two A components, lethal factor (LF) and edema factor (EF), are the intracellularly active enzymes. LF is a Zn-metalloprotease that cleaves MAP kinase kinases [22,23] and Nlrp1 [24]. EF is a Ca^{2+} and calmodulin-activated adenylyl cyclase [25,26]. Protective antigen (PA), named this way for its ability to elicit protective antibodies (the property utilized in the anthrax vaccines), is a receptor for LF/EF binding, which mediates their translocation. The anthrax toxin intracellular delivery occurs in several stages. After binding to its cellular CMG2 and/or TEM8 receptors and proteolytic cleavage to PA_{63}, by extracellular furin, PA oligomerizes on the host cell surface to form heptameric [27] and/or oligomeric ring-shaped pre-pores [28,29] creating three [30] or four [28] LF and/or EF binding sites. After receptor-mediated endocytosis [31], the anthrax toxin AB complexes are delivered to the acidic environment of the early endosome. There, the PA oligomers undergo substantial conformational changes leading to their insertion in endosomal limiting, and possibly in intraluminal vesicle membranes [32], eventually forming an extended 180-Å long "flower-on-a-stem" cation-selective [33] channel with a 75 Å long bud and a 105 Å long stem and radius varying from 16 Å to 3.5 Å [34]. This channel is generally believed to work as an effective translocase that unfolds and allows for translocation of LF and EF into the cytosol under a pH gradient across the late endosomal limiting membrane ($pH_{endosome} < pH_{cytosol}$) [35,36]. An alternative model suggests that the anthrax toxin catalyzes the rupture of the endosomal membranes, which leads to the consequent delivery of the toxin complexes into the cytosol [37].

2. Results

2.1. Two Modes of G2-NH₂ PAMAM Dendrimer Inhibition of PA₆₃ Channel

Figure 2 illustrates the bimodal effect of G2-NH₂ PAMAM *cis*-solution addition on the ionic current through a single PA_{63} channel incorporated into planar lipid bilayer membranes. To obtain reliable statistics on G2-NH₂/PA₆₃ interaction, we performed the single channel measurements in 1 M KCl. A decrease in salt concentration led to a dramatic increase in the blocker lifetimes, suggesting the involvement of the long-range Coulomb interactions. Quantitative analysis of the process at lower, e.g., physiological salt concentrations, proved impossible over the course of our experiments. Previously, the PAMAM dendrimers were reported to be ~100–900 times more effective when added to the *cis*-side of the membrane, which is also the side of PA_{63} addition [5]. PA_{63} insertion was shown to be almost exclusively unidirectional [2,9,33], with the bud, LF/EF binding part of the channel, facing the *cis*-side solution (corresponding to the endosome interior), and the stem part facing the *trans*-side solution (corresponding to the cytosol or ILV interior). The single channel current recordings show that, in a manner similar to the cationic β-cyclodextrin [14] and G1-NH₂ PAMAM dendrimer blockers [5], the G2-NH₂ inhibitive action is bimodal (Figure 2A). Firstly, G2-NH₂ addition generates complete but reversible blockages of ion current through a single channel (marked with two blue ovals, Figure 2A). Frequency of these events increases in a concentration-dependent manner and is a strong function of the applied transmembrane voltage (Figure 2B). Note: *cis*-positive sign of the applied transmembrane voltages corresponds to the inside-positive voltage gradient across endosomal limiting membranes. Secondly, G2-NH₂ addition led to a dramatic increase in the voltage-dependent gating of PA_{63} channels (Figure S2), seen as prolonged closing events (marked with red ovals, Figure 2A, middle and right). Higher concentrations of G2-NH₂ and higher voltages compared to the ones reported earlier ($K_D = (7.2 \pm 4.7) \times 10^{-9}$ M at $V = 20$ mV) in the multichannel systems were needed because of the increased supporting electrolyte concentrations (1 M vs. 0.1 M) that, by electrostatically screening charges on both the blocker and the channel, weakened blocker binding.

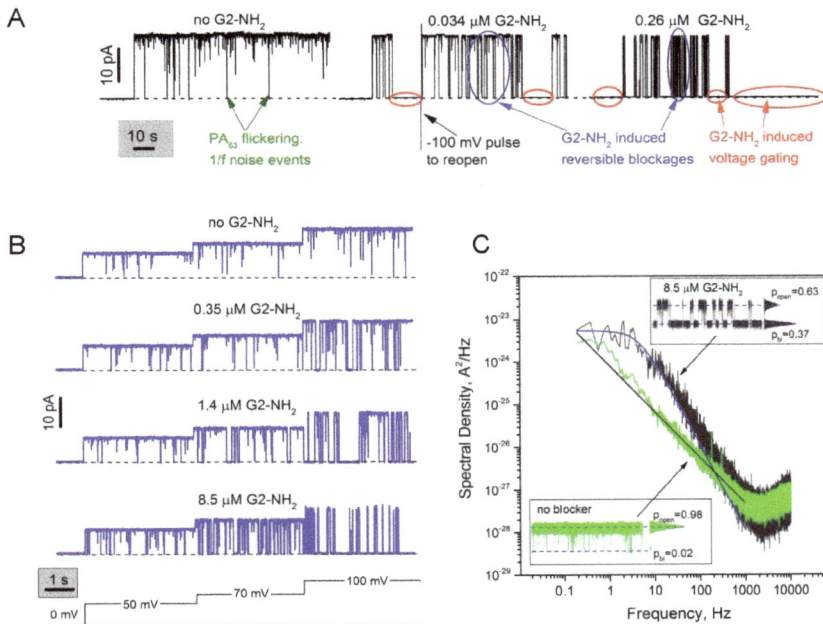

Figure 2. Modulation of a single PA_{63} channel current by $G2-NH_2$ PAMAM dendrimer. (**A**) Two modes of $G2-NH_2$ action on a single PA_{63} channel. In the absence of $G2-NH_2$ (left), PA_{63} remains in an open state. Fast flickering between the open and closed states (the so-called $1/f$ noise) is mostly but not completely removed by averaging over a time interval of 50 ms. In the presence of two different $G2-NH_2$ concentrations (middle and right), both blocker-induced reversible blockage and prolonged voltage gating events are seen. Recordings were taken at 100 mV applied voltage; (**B**) First mode of action. In the absence of the blocker (top) the ion movement is determined by the geometry and surface properties of the PA_{63} pore. In the presence of $G2-NH_2$ in the *cis* compartment of the bilayer chamber (three following rows), the channel gets reversibly blocked. At higher concentrations of $G2-NH_2$ (bottom) the blockages, which are seen as downward spikes, are more frequent. The probability of finding PA_{63} in the blocked state increases with *cis*-positive transmembrane voltage increase (50, 70, and 100 mV are shown). Current tracks were averaged over a time interval of 2 ms; (**C**) Power spectral densities of the $G2-NH_2$–induced PA_{63} current fluctuations (spectrum in grey) can be fitted by a single Lorentzian at frequencies of <1000 Hz in contrast to $1/f$ noise in the absence of $G2-NH_2$ (spectrum in green). A clear deviation from a single Lorentzian dependence at $f > 100$ Hz could to some degree be explained by the $1/f$ noise and the partial dendrimer degradation and breakage with formation of the imperfect cationic substrates, capable of blocking PA_{63} channels with shorter lifetimes (mass spectra and NMR characterization is given in Figures S5 and S6). Inserts: 1-s current tracks averaged over a time interval of 0.2 ms for blocker-free (bottom left) and 8.5 µM $G2-NH_2$ (upper right) recordings. All-point histograms are shown to the right of the recordings; p_{open} and p_{bl} denote the probability of the PA_{63} channel being in conductive and non-conductive states, respectively. Applied voltage was 100 mV; dashed lines represent zero current levels.

The fast reversible current fluctuation induced by $G2-NH_2$ in the parts of current tracks with excluded voltage-dependent gating can be described as a two-state memoryless Markov process, where both the residence time in the blocked state and the channel lifetime in the unblocked state (the time between blockages) are described by exponential distributions. This is demonstrated by the Lorentzian shape of the power spectral density of $G2-NH_2$-induced current fluctuations at $f < 1000$ Hz (Figure 2C, spectrum in grey fitted by the smooth blue solid line through the experimental curve). This relatively straightforward kinetic analysis was to a certain extent complicated by a number of factors, namely the two types of complex non-Markovian channel gating described in detail previously [14,15]. The first type of gating is induced by the applied voltage that brings the PA_{63} channel into a nonconductive state, which seems to be characteristic for β-barrel channels in general [38]. This voltage dependent gating was especially prominent at *cis*-side negative voltages [33]; thus applied voltages as low as − (10–20) mV led to the prolonged channel closures. The fact that the β-barrel PA_{63} channel tends to stay closed when positioned under non-physiological inside-negative voltage gradients, adds fuel to the little rusty but still very interesting debate about the significance [39] and mechanism [40] of voltage gating for the unconventional channel [41] function. Note that while some researchers show that the voltage-dependent β-barrel channel closure represents nothing more than an artifact of bilayer lipid experiments [42], others report clear evidence of physiological significance of voltage gating in β-barrel channels [43,44]. In one way or another, this circumstance has largely limited our ability to collect reliable statistics on channel/blocker binding reaction at negative and high positive voltages, especially because in many cases the dendrimer addition has significantly enhanced the voltage sensitivity of the channel. The second type of PA_{63} non-Markov gating is the so-called voltage-independent fast flickering $1/f$ noise between the open and completely closed states that was earlier described as a universal intrinsic property of the pore-forming components of AB type toxins, PA_{63}, C2IIa, and Ib, both at the single [2,3,14,15] and multi-channel [8,9] level. The current noise power spectrum of the non-modified PA_{63} channel contains a $1/f$-like voltage-independent [14,15] component (Figure 2C, see the spectrum in green and the corresponding current track (left insert) shown at 0.2 ms time resolution). Even though the $1/f$ flickering is not among the immediate points of interests of the current publication, the universality of the $1/f$ flickering and the fact that the F427A mutant of PA_{63}, which lacks the ϕ-clamp [7] and therefore A-component translocation functionality, was devoid of the $1/f$ noise behavior, deserves to be studied more closely. Within the limits of this study, we had to be very careful to uncouple the $1/f$ fast-flickering events and the dendrimer-induced reversible blockages, especially under conditions where the closed time distributions of these events partially overlap. For example, in Figure 2A (left) we show several relatively long $1/f$ flickering events that are still seen even at low, 50 ms time resolution. To quantify kinetic parameters of the G2-NH$_2$-induced reversible blockages, we primarily used current noise spectral analysis (Figure 2C) instead of the direct counting of open and closed event durations. The direct counting approach does not allow us to distinguish between open times of the $1/f$ noise closures and the dendrimer-induced reversible blockages, because a combined open time distribution for these two processes was single-exponential, which may be explained by the fact that there is only one open state of the channel. The average lifetime of G2-NH$_2$ in the channel pore (t_{res}) and average time between blockages (t_{on}) were found correspondingly as $t_{res} = 1/2\pi f_c (1 - p_{bl})$ and $t_{on} = 1/2\pi f_c p_{bl}$, where f_c is the corner frequency of Lorentzian and p_{bl} is the probability of finding PA_{63} in the blocked state [45]. Ideally, the probability of the channels being in the blocked state could be directly determined as $p_{bl} = \frac{I_0 - I_{ave}}{I_0}$, where I_0 is the ion current through the completely open channel, and I_{ave} is the average ion channel current modified by the blocker. However here, to account for the $1/f$ fast flickering, the equation was corrected, assuming independence of these two processes as follows: $p_{bl} = \frac{I_{ave}^{free} - I_{ave}}{I_{ave}^{free}}$, where I_{ave}^{free} is the average current through the PA_{63} channel measured in blocker-free solutions [14]. Note: to determine I_{ave}^{free} and I_{ave}, the prolonged voltage gating closures (both intrinsic and dendrimer-induced) were excluded from the open and closed states probability analysis.

The second mode of dendrimer-induced current inhibition was hard to describe quantitatively because the blocker-induced channel closures often appeared to be irreversible, lasting for minutes or longer. To reopen the channel, we either had to apply 0 mV or to reverse the voltage sign (shown in Figure 2A, middle) which did not allow us to collect reliable statistics on kinetic parameters of the second mode of channel blockage. Moreover, as described above, the voltage-induced closures were also recorded in the absence of blocker, and voltage-sensitivity and probability of finding a channel in the closed state varied from channel to channel. However qualitatively, this process evinced all the key characteristics of the voltage-induced gating of β-barrel channels, such as strong voltage dependence, prolonged closures (minutes), and difficulties in reopening channels even when voltage was reduced to zero. Paradoxically, channel reopening was often possible with abrupt second-long pulses of high voltages of opposite sign (marked in Figure 2A, middle track), that, in turn, also induced voltage dependent closures if applied longer.

2.2. Role of PAMAM Dendrimer Surface Chemistry in PA$_{63}$ Blockage

The common mechanism of inhibiting cation-selective PA$_{63}$ involves blockage of the channel's lumen by positively charged molecules [2,3,5,7,9,10,14,33,46,47]. However, on a multichannel level in 0.1 M KCl, it has been demonstrated that G2 and G3 PAMAM dendrimers functionalized with surface hydroxyl groups (G2-OH and G3-OH) inhibited PA$_{63}$ channels in a concentration-dependent manner [5]. The *IC*$_{50}$ values were about 20 (Table 1) and 9 times lower compared to those of G2-NH$_2$ and G3-NH$_2$, respectively. Typical multichannel titration curve analyses are given in Figure S3. Interestingly, even in the presence of PAMAM dendrimer functionalized with negatively charged carboxyl surface group (G1-COONa, but named G0.5 by the manufacturer), some small current decrease was recorded, though compound activity was not high enough to reach 50% inhibitory concentrations. In order to understand how the PA$_{63}$ channel selects among the PAMAM dendrimer blockers, here we investigated kinetic parameters of the binding reaction directly comparing G2-NH$_2$, G2-OH, G2-SA, and G2-COONa PAMAM dendrimers binding to PA$_{63}$ on a single channel level (Figure 3 and Figure S4). As before, the single channel measurements are performed in 1 M KCl (Figure 3).

Table 1. Inhibition of PA$_{63}$ channel ion current by intact G2 PAMAM dendrimers and G3 PAMAM dendron expressed as experimental *IC*$_{50}$ values.

	PA$_{63}$/PAMAM Dendrimer Binding Reaction, *IC*$_{50}$					
	G2-NH$_2$	G2-OH	G2-SA	G2-COONa	G2 75% OH 25% NH$_2$	G3-NH$_2$ Dendron
0.1 M KCl	7.2 ± 4.7 nM	142 ± 36 nM	879 ± 50 μM	>14 mM	122 ± 35 nM	16.4 ± 4.0 nM
1 M KCl	5.1 ± 2.6 mM	>30 mM	1.7 ± 0.2 mM	not determined	7.7 ± 0.2 mM	7.8 ± 1.0 mM

All data were calculated as means from at least two separate multichannel experiments; the errors are standard deviations. 0.1 M and 1 M KCl solutions at pH 6 were buffered by 5 mM MES. Recordings were taken at 20 mV applied voltage, which was *cis*-side positive. Note: because G2-COONa activity in 0.1 M KCl was too low to reliably detect *IC*$_{50}$, we did not attempt to perform experiments in 1 M KCl bathing solutions, where charges on both the channel and the dendrimer were to a certain extent screened by the bathing solution counterions.

Figure 3A gives four representative recordings of a single PA$_{63}$ channel current modified by different concentrations of G2-NH$_2$, G2-OH, G2-SA (three lower rows) compared to dendrimer-free solution (top). When added to the *cis* compartment solutions, all dendrimers reversibly blocked the channel, with *cis*-positive transmembrane voltages significantly enhancing the blockage with G2-NH$_2$ and G2-OH, but not G2-SA (Figure 3B,C). Two modes of G2-NH$_2$-induced current blockage are evident at a very low sub-μM blocker concentration at 100 mV applied voltage. In the presence of 0.35 μM of G2-NH$_2$, probability of finding the channel in the closed state increases to 35% (second row) compared with 0.3% in the dendrimer-free solution (first row). G2-OH dendrimer addition (third row) causes similar complete but reversible channel closures (first mode of blockage), however more than 2000 times higher blocker concentrations were required to increase probability of the closed state to 20%. At voltages ≥100 mV, we also detected the second mode of channel closure (voltage gating events), however this effect was significantly less pronounced compared to the one observed for G2-NH$_2$. At comparable sub-mM concentrations and 100 mV, G2-SA, modified with negatively charged surface groups, (fourth row) shows only weak interaction with probability of closed state equal to 2%. We could not detect any concentration-dependent channel fluctuations upon addition of

G2-COONa to the *cis* compartment of the bilayer chamber. Instead, high concentration of G2-COONa (3 mM) caused membrane instability and, eventually, breakage (Figure S4).

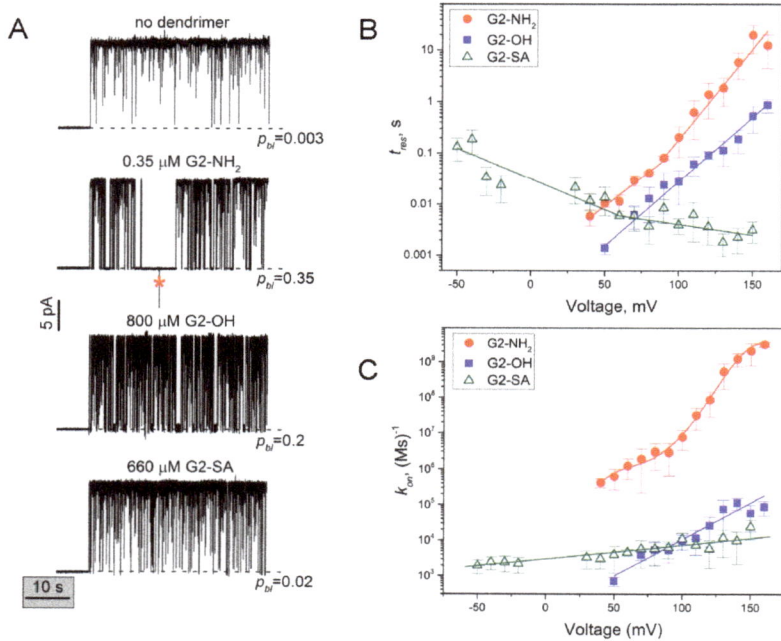

Figure 3. Influence of the PAMAM dendrimer surface chemistry on the PA$_{63}$ single channel current inhibition. (**A**) PA$_{63}$ channel current tracks in the absence (top) and presence of G2-NH$_2$, G2-OH, and G2-SA PAMAM dendrimers in the *cis* side of the bilayer chamber (three lower rows). In the dendrimer-free solution, the channel remains open with only 0.3% probability of being in the closed state (p_{bl}) due to intrinsic $1/f$ noise flickering (mostly filtered). Addition of G2-NH$_2$ (second row) induces two modes of PA$_{63}$ current blockage, fast reversible fluctuations and longer voltage gating type of closures (marked by "*"). At 0.35 µM G2-NH$_2$, the probability of finding the channel in the closed state increases to 35%. The addition of 800 µM G2-OH (third row) causes fast reversible blockages, whereas the voltage gating events are less pronounced; the probability of finding PA$_{63}$ in the closed state is 20%. G2-SA addition (bottom row) causes reversible blockages at similar sub-mM concentrations, with a 2% probability of finding PA$_{63}$ in the closed state. Applied voltage was 100 mV; current tracks were averaged over a time interval of 10 ms. The dashed lines represent zero current levels; (**B**) Residence times of dendrimer binding reaction plotted as functions of transmembrane voltage. While residence time of G2-OH (filled squares) increases exponentially with voltage (solid line through the data points), residence times of G2-NH$_2$ (filled circles) and G2-SA (open triangles) show more complex progression, increasing and decreasing with the voltage increase respectively. The data were split into two voltage intervals and fitted with two separate exponents (solid lines) (**C**) On-rate of dendrimer blockage defined as the inverse of the average open channel life time as a function of voltage for G2-NH$_2$ (filled circles), G2-OH (filled squares), and G2-SA (open triangles) blockers. Note the voltage sensitivity of k_{on} for all three dendrimers and the significantly higher absolute values of the on-rate constant in the case of G2-NH$_2$.

2.3. The Rate Constants of Dendrimer's First Mode of Binding Reaction are Voltage Dependent

For all three blockers, the on- and off-rate constants of the PA_{63}/dendrimer binding reaction varied with the applied voltage (Figure 3B,C). The shown data were obtained with power spectral analysis of the reversible current fluctuations, using the fitting by single Lorentzian spectra as described above (Figure 2C). For voltages ≥ 130 mV, when strong blocker-induced voltage gating (second type of blockage) prevented us from collecting long current recordings suitable for the power spectral analysis, the data were analyzed by averaging over direct measurements of blocked and open event durations. The on-rate was calculated as $k_{on} = 1/t_{on}c_{PAMAM}$, where t_{on} is time between successful blockages, and c_{PAMAM} is dendrimer bulk concentration. While in the case of G2-OH, the dendrimer with the positively charged tertiary amine interior and neutral hydroxyl surface groups, t_{res} was shown to increase exponentially (linear fit in the semi-logarithmic scale in Figure 3B), in the case of the "highly" cationic G2-NH$_2$ binding, we observed non-exponential voltage dependence of t_{res} (fitted with two linear dependences in Figure 3B). The binding time in the presence of G2-SA, the dendrimer that is made of the positively charged PAMAM core and 16 negatively charged succinamate surface groups, was shown to have inverse voltage dependence, with the t_{res} binding time decreasing with *cis*-positive voltage decrease. Figure 3C shows the on-rate dependence of the PAMAM dendrimer/binding reaction on the applied voltage. Interestingly, for all three dendrimers we observed a voltage-dependent increase in the binding reaction rate constants (Figure 3C), which indicates that high voltages make PAMAM dendrimer capture by the PA_{63} easier. At the same time, k_{on}, and, therefore the number of individual blockage events, was drastically higher for G2-NH$_2$ compared with both G2-OH and G2-SA.

2.4. PA_{63} Blockage by Imperfect Cationic PAMAM Dendrimers

Activity of PAMAM dendrimers was previously shown to increase dramatically when they were degraded at the amide linkage, to a heterodisperse population of dendrimers of different molecular weights [20]. These less sterically constrained and hence, more flexible, "imperfect" dendrimers were reported to show significantly enhanced transfection activity compared to the intact dendrimers. To test if a similar effect could be achieved with the pore blockage, we previously investigated the channel blocking activity of two different types of "imperfect" PAMAM dendrimers. First, we tested a mixed-surface G2 75% OH 25% NH$_2$ charge-dispersed PAMAM dendrimer, where the proportion of the positively charged surface amino groups was only 25% on average. In 0.1 M KCl, this dendrimer was about 17 times less active against PA_{63} compared to G2-NH$_2$. A notable increase in the blocker's activity was achieved with the structurally incomplete 8+ charged G2 dendron, which was about 26 times more active, compared to the 8+ charged G1-NH$_2$ dendrimer [5]. Here we investigate the G2 75% OH 25% NH$_2$ dendrimer activity and also use the 16+ charged G3-NH$_2$ dendron to directly compare their activity with the 16+ charged G2-NH$_2$ dendrimer on a single-channel level (Figure 4). In a manner similar to G2-NH$_2$ intact dendrimers, the inhibitive action of the G2 75% OH 25% NH$_2$ dendrimer and G3-NH$_2$ dendron was bimodal. Both the short reversible blockage events and the prolonged closures were detected after the addition of sub-µM concentrations of either of these two imperfect dendrimers to the *cis*-compartment solutions (Figure 4A). While the effective concentrations were comparable to those used for G2-NH$_2$ (and significantly lower compared to G2-OH and G2-SA), the voltage sensitivity of the binding reaction on-rates was substantially different (Figure 4B,C).

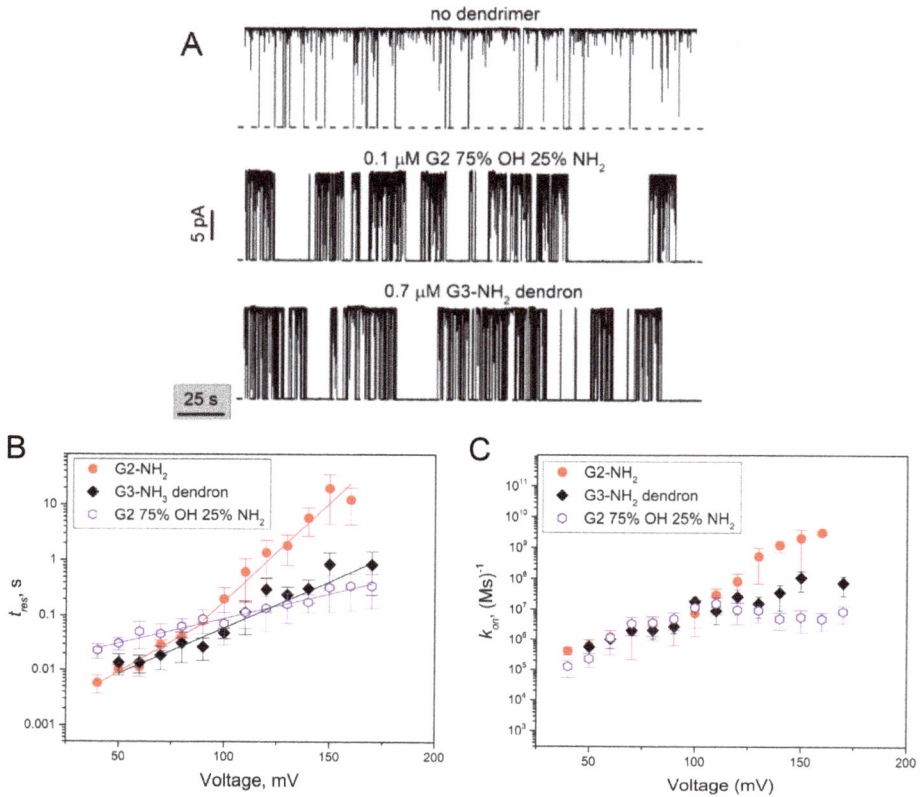

Figure 4. Modulation of a single PA_{63} channel current by imperfect G2-75% OH 25% NH_2 dendrimer and G3-NH_2 dendron. (**A**) The sub-μM addition of G2 75% OH 25% NH_2 dendrimer (middle) and G3-NH_2 dendron (bottom) to the *cis*-side of the bilayer chamber results in PA_{63} current inhibition. Two modes of the blocker action are clearly seen. In the absence of a blocker (top), the channel is open; the intrinsic $1/f$ flickering events are mostly filtered by averaging over a time interval of 50 ms. Applied voltage was 100 mV, measurements were performed in 1 M KCl. The dashed lines represent zero current levels; (**B**) Residence time of imperfect G2 75% OH 25% NH_2 dendrimer (open hexagons) and G3 dendron (filled diamonds) binding increases exponentially with voltage increase. Residence time for G2-NH_2 dendrimer (filled circles) is as discussed in Figure 3B; (**C**) On-rate constant as a function of voltage increases and then shows saturation at $V > 100$ mV in the case of G2 75% OH 25% NH_2 dendrimer and G3 dendron, but not with G2-NH_2 where it continues increasing.

3. Discussion

3.1. Two Modes of G2-PAMAM Dendrimer Inhibition of PA_{63} Channel

In this paper, we used the single channel planar lipid bilayer technique to present evidence that second generation PAMAM dendrimers inhibit the PA_{63} channel of anthrax toxin by ion current blockage. For all tested G2 PAMAM dendrimers and the G3-NH_2 PAMAM dendron, the channel current inhibition was bimodal. The first mode of the ion current inhibition was observed in the form of complete (100% of total channel conductance) but reversible ion current fluctuations that were described by a two-state Markov process, with one state being an open, dendrimer-free state and second state being a blocked, dendrimer-bound state. Interestingly, not only the three tested cationic blockers (intact G2-NH$_2$, mixed surface 75% OH 25% NH$_2$ dendrimers, and G3-NH$_2$ dendron) but also G2-OH and G2-SA dendrimers (functionalized, respectively, with neutral and negatively charged terminal groups) reversibly blocked the K$^+$ current through PA$_{63}$, apparently physically entering the channel's permeation pathway. Figure 5 summarizes the equilibrium dissociation constants for all five blockers that were calculated using the blocker/channel binding reaction kinetic constants shown in Figures 3 and 4 as $K_D = \frac{k_{off}}{k_{on}}$, where $k_{off} = \frac{1}{t_{res}}$. It can be seen (Figure 5) that G2-NH$_2$ activity significantly exceeds (lower K_D values) that of G2-OH and G2-SA, which is mostly determined by the dramatically higher on-rates of the G2-OH/PA$_{63}$ and G2-SA/PA$_{63}$ binding reactions rather than by the reactions' off-rates. Thus, under 50 mV of applied voltage, t_{res} of G2-OH was about 10 times lower, and t_{res} of G2-SA was comparable with that of G2-NH$_2$ (Figure 3B). The on-rates were about three-fold and two-and-a-half-fold higher for G2-NH$_2$ compared with G2-SA and G2-OH, respectively (Figure 3C). In contrast, the on-rate of the binding reaction was nearly identical when G2 75% OH 25% NH$_2$ dendrimer and G3-NH$_2$ dendron were investigated and compared with G2-NH$_2$ dendrimer at 50 mV (Figure 4C).

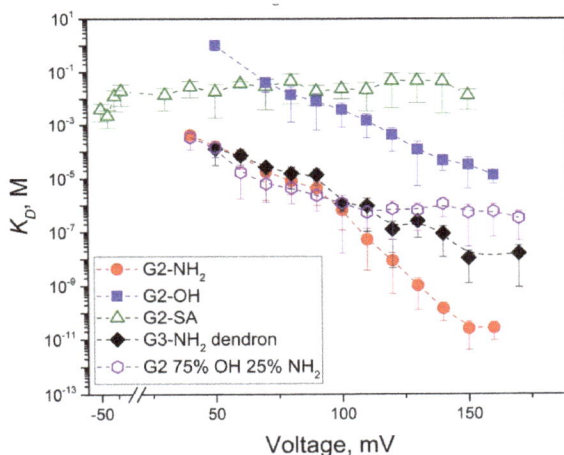

Figure 5. Equilibrium dissociation constants of the PAMAM dendrimer binding reaction to a single PA$_{63}$ pore plotted as a function of applied transmembrane voltage. Note that only first mode of blockage type events were considered. K_D values significantly decrease with voltage (stronger binding) for all the shown blockers, except G2-SA. The G2-OH dendrimer K_D values are significantly higher (less effective binding) compared to those for G2-NH$_2$, G2 75% OH 25% NH$_2$ dendrimers and the G3-NH$_2$ dendron. The experiments were performed in 1 M KCl at pH 6.

In addition to the reversible two-state Markov process blockage events, all five blockers caused long voltage-dependent PA$_{63}$ closures that, in many aspects, resembled the typical characteristics of the classical voltage gating of β-barrel ion channels [42] (Figure S2). The quasi-irreversible character of the voltage gating allowed us to perform only qualitative analysis of this type of blocker-induced closure. In a manner similar to the reversible Markov PA$_{63}$ blockage process, the voltage gating type of closures appeared to be stronger when the dendrimers were decorated with positively-charged terminal groups and was very weak in the case of G2-SA (see e.g., Figure 3A where no voltage dependent closures are seen in the presence of 660 μL of G2-SA over ~50 s recording). Previously, we described the two similar modes of the blocker-induced channel inhibition with the 7-fold symmetrical β-cyclodextrin inhibitors [14,15]. Because the molecular mechanism, possibly a universal one, of the voltage gating observed for many functionally distinct β-barrel channels in planar lipid bilayers has yet to be clarified [40] and its physiological relevance has been called in question [39], we cannot make a strong judgement on the biological importance of the second mode of the dendrimer-induced PA_{63} current inhibition. However, we want to emphasize that when we attempted to study the dendrimer and cyclodextrin binding reaction single-channel kinetics at low, close to physiological salt concentrations (e.g., in 0.1 M solutions), the second type of the current inhibition was very strong. This led to prolonged and complete channel closures at very low, (20–50) mV applied voltages, precluding us from performing kinetic investigation of the PA_{63}/multivalent blocker binding reaction. For the same reason, we had difficulties studying the first type of the blockages at voltages >(150–180) mV, because the quasi-irreversible voltage gating events were observed at extremely low blocker concentrations (0.01–0.1 nM) when the number of the first type of reversible blockage events was too low compared to the natural $1/f$ current fluctuations of this very complex channel. Previously we reported a reasonably good linear correlation ($R = 0.84$) between activity of the PA_{63} channel cyclodextrin inhibitors in RAW 264.7 cells and in the multichannel reconstitution assays [48], allowing us to suggest that the second voltage gating type of closures is physiologically relevant. Moreover, the strong asymmetry that we and others observed in non-modified PA_{63} voltage gating with the voltage sign (the channel tends to close even at very low *cis*-negative voltages) [33,49], suggests that the voltage gating could be an internal tool for PA_{63} to stay closed when being occasionally inserted into off-target bilayers.

Note that the K_D data in Figure 5, when approximated to 20 mV, are higher compared with the equilibrium IC_{50} values measured at 20 mV (Table 1). All the blockers appear to be more effective when used at the multichannel level. The explanation of this apparent discrepancy is simple and identical to the one previously suggested for the PA_{63}/7+β-CD binding reaction [14]. The IC_{50} values measured at the multichannel level contain both the fast reversible dendrimer-induced blockage events (first mode of dendrimer action) and the prolonged voltage gating closures (second mode of action), whereas the voltage gating events were intentionally excluded from binding reaction kinetics analysis at the single channel level. As discussed before, we believe that the second mode of PA_{63} current inhibition is related to the voltage gating of β-barrel channels, well-known to any electrophysiologist who works with these channels. It appears that β-CD and dendrimer addition significantly lowers the so-called "critical voltage" needed for the channel gating. Because we currently lack understanding of the voltage gating mechanism, we cannot comment on the physiological relevance of this second mode of current blockage (frequently a more intense one) in anthrax toxin inhibition. Note that the β-CD and dendrimer blockers were previously shown to effectively protect cells [2,3] and animals [50] against the anthrax toxin.

3.2. Voltage Dependence of the Reversible Dendrimer/PA63 Interaction

With all previously tested β-cyclodextrin PA_{63} blockers, nearly exponential (linear in semi logarithmic scale) voltage dependence of t_{res} was reported with almost identical voltage sensitivity between two tested 7+ charged β-CD blockers, whereas the k_{on} values were only weakly voltage-dependent [14,15]. Moreover, it was the off-rate and not the on-rate that mainly determined the earlier reported difference in 7+β-CD potency. Details of the mechanism of the PAMAM dendrimer-induced PA_{63} current blockage are different. Thus in the case of G2-OH, the dendrimer with a positively charged tertiary amine interior and neutral hydroxyl surface groups, t_{res} was shown to increase exponentially with voltage (linear fit in the semi-logarithmic scale in Figure 3B, filled squares) with the slope of the logarithm of the lifetimes versus voltage dependence $dlgt_{res}/dV = (25 \pm 0.6) \times 10^{-3}$ $(mV)^{-1}$. In contrast, with the "highly" cationic (16 terminal amino groups) G2-NH$_2$, we observed a non-exponential t_{res} increase with voltage. The t_{res} voltage dependence data were broken into two intervals and each was fitted with a single exponent (shown as two linear dependences in the semi-logarithmic scale in Figure 3B, filled circles). The slopes of the logarithm of the lifetimes versus voltage dependence for low and high voltages were: $dlgt_{res}/dV = (22 \pm 2) \times 10^{-3}$ $(mV)^{-1}$ at low *cis*-positive voltages and $dlgt_{res}/dV = (35 \pm 2) \times 10^{-3}$ $(mV)^{-1}$ at high *cis*-positive voltages. The binding time in the presence of G2-SA, the dendrimer that is made of the positively charged

PAMAM core and 16 negatively charged succinamate surface groups, was shown to have inverse voltage dependence, with the t_{res} decreasing with *cis*-positive voltage decrease. The G2-SA residence time voltage dependence was also approximated with two separate single exponential dependences (Figure 3B, open triangles), one at lower voltages ($\mathrm{dlg}t_{res}/\mathrm{d}V = -(12 \pm 1) \times 10^{-3}$ $(\mathrm{mV})^{-1}$) and another at higher voltages ($\mathrm{dlg}t_{res}/\mathrm{d}V = -(4.2 \pm 0.9) \times 10^{-3}$ $(\mathrm{mV})^{-1}$). It appears that the high *cis*-positive and high *cis*-negative applied voltages increase voltage sensitivity of the $PA_{63}/G2\text{-}NH_2$ and $PA_{63}/G2\text{-}SA$ binding reactions, respectively. Thus, the t_{res}/voltage dependence data could be used to determine the so-called effective "gating charge" [51], a parameter characterizing sensitivity of the blockage reaction to voltage [52], as $\delta z = k_B T \frac{\mathrm{d}2.303\mathrm{lg}t_{res}}{\mathrm{d}V}$. Here, δ is the dimensionless "apparent electrical distance" to the blocking site, z is the blocker charge, V is the applied voltage, and k_B and T have their usual meaning as the Boltzmann constant and absolute temperature (in degrees Kelvin), respectively. The G2-OH binding could be characterized by $\delta z = 1.50 \pm 0.03$ elementary charges, the G2-NH$_2$ binding by $\delta z = 1.3 \pm 0.2$ (low voltage) and $\delta z = 2.10 \pm 0.07$ (high voltage) elementary charges, and G2-SA binding by $\delta z = -0.72 \pm 0.08$ (low voltage) and $\delta z = -0.25 \pm 0.05$ (high voltage) elementary charges. The twisted residence time voltage dependence observed with G2-NH$_2$ and G2-SA but not with G2-OH could tentatively be explained by structural reorientation of the terminal amino and succinamate groups or solvation structure (shell water and counterions) reorganization under the applied electrostatic field. Interestingly, k_{on} increases as a function of voltage (Figure 3C) showing that high voltages facilitate dendrimer delivery to the binding site, in contrast to the very weak voltage dependence of the binding reaction on-rate reported earlier for 7+β-CDs but in a manner similar to that reported for alpha-synuclein/α-hemolysin binding reaction [53]. Surprisingly, we also observed a slight increase in the on-rate with *cis*-positive voltage increase even for G2-SA dendrimer, which carries 16 negatively-charged terminal groups.

We have also investigated the role of the so-called "imperfect" dendrimers in the PA$_{63}$ blockage dynamics using G2 75% OH 25% NH$_2$ dendrimer which, on average, has only four terminal positive charges and G3-NH$_2$ dendron decorated with 16 positively-charged terminal groups. The residence times of the blockers inside the channel changed exponentially with voltage showing more shallow slopes compared with those earlier discussed for G2-NH$_2$: $\mathrm{dlg}t_{res}/\mathrm{d}V = (16.8 \pm 1.5) \times 10^{-3}$ $(\mathrm{mV})^{-1}$ for the G3-NH$_2$ dendron and $\mathrm{dlg}t_{res}/\mathrm{d}V = (9.1 \pm 0.3) \times 10^{-3}$ $(\mathrm{mV})^{-1}$ for G2 75% OH 25% NH$_2$ dendrimer (Figure 4B). These two systems can be characterized by the effective "gating charge" $\delta z = 0.92 \pm 0.16$ (G2-NH$_2$ dendron) and $\delta z = 0.53 \pm 0.03$ (G2 75% OH 25% NH$_2$) elementary charges showing weaker residence time voltage sensitivity compared to G2-NH$_2$. At the same time at voltages <80 mV, G3-NH$_2$ dendron lifetime is comparable and G2 75% OH 25% NH$_2$ lifetime is even higher than that for G2-NH$_2$ (Figure 4C). At $V < 120$ mV the binding reaction on-rates are comparable for all three blockers, however at $V > 120$, t_{on} shows only weak voltage dependence for both the imperfect dendrimers.

4. Conclusions

Since their discovery in 1985 by Tomalia [54,55], PAMAM dendrimers have been the subject of thorough theoretical and experimental investigation in soft-matter physics, not only due to their seemingly high commercial potential but also because of their unique "ultrasoft colloid" properties bridging the gap between polymers and hard spheres [56]. PAMAM dendrimer conformational flexibility in solutions and the effect of solvent and pH on their structure, swelling, charge, counterion distribution, degree of protonation, and deformability have been addressed in a significant number of publications [57–71]. At the same time, single channel studies investigating specifics of the dendrimer dynamics in ion channel confinement are limited. In 2000, a rapid nuclear pore sizing patch-clamp method based on the calibrated fluorescently-labeled amino-terminated dendrimers was described [72]. In 2007, sulfhydryl-reactive poly(amido amine) G2, G3, and G5 dendrimers of second, third and fifth generations decorated with a mixed surface of terminal hydroxyl and amine groups were designed to interact with α-hemolysin channels that contained engineered cysteine residues [73], with the ultimate goal of modifying the stochastic-sensing properties of α-hemolysin upon addition of the charged and dense dendrimers into its lumen. The dendrimers acted as both an ion-selectivity filter and a molecular sieve, regulating the passage of small- and macromolecules. In 2013, polypropylenimine dotriaconta-amine G3 and G4 dendrimers were tested against the *E. coli* E69 pore-forming Wza K30 capsular polysaccharide transporter; however, no detectable inhibitory activity was reported [74]. In 2014, PAMAM dendrimers were reported to effectively block the ion-channel forming components of the anthrax and C2 toxins using planar lipid bilayer measurements and cell assays [5]. More recently, α-hemolysin/PAMAM dendrimer system was used to investigate the molecular process of ion channel confinement and its effect on dendrimer conformation using single channel measurements and molecular dynamics simulations [75]. The authors have shown that the electrophoretic migration of the polycationic dendrimers into a confined space is determined by the generation-dependent compressibility of the dendrimers rather than by their diameter. The ion channel nanoscale confinement had also reduced the PAMAM dendrimer protonation. Just recently, fully atomistic molecular dynamics simulations were used to investigate pH-dependent blockage (the authors call it "gating") of the cytolysin A pore by PAMAM dendrimers [76]. The protonated dendrimers were able to adapt a more extended conformation, effectively blocking about 91% of the channel current, whereas the non-protonated dendrimers were more compact, which created some void space for water and ion passage and about 31% reduction in current.

In this study, we investigated the effect of the PAMAM dendrimer surface chemistry and structural integrity on their ability to enter and block the ion (and probably lethal and edema factor) translocation pathway of the anthrax toxin channel. Considering the electrostatic nature of the cationic blocker interaction with the strongly cation-selective PA_{63} pore, it comes as no surprise that the residence time of the dendrimer/channel binding reaction turned out to be dependent on both blocker chemistry and transmembrane voltage. One of the unforeseen findings made in this study is the increase in the dendrimer capture rate (the binding reaction on-rate) with the *trans*-negative voltage increase and the strong on-rate dependence on dendrimer surface chemistry. Indeed, the on-rate, which is proportional to the number of effective binding events, is traditionally believed to be determined by a correlation between the particle size and the channel entry area (or squared radius). The applied electrical field that falls primarily across the channel (traditionally, across the membrane) is not expected to significantly influence the capture kinetics at distances greater than 3 Å away (1 Debye length in 1 M KCl) from the channel entrance. Thus, previously reported strong voltage sensitivity of a particle capture for the tubulin/VDAC [77] and alpha-synuclein/α-hemolysin [53] binding reactions was explained by the tubulin and alpha-synuclein interaction with the bilayer lipid membranes as an essential first step before the channel lumen binding. This explanation is clearly not appropriate for dendrimer/PA_{63} binding because of its compact nature. The size of the blocker (29 Å) and extremely elongated structure of the PA_{63} channel [34] means that it may extend to a distance greater than 100 Å above the membrane surface. Nevertheless, PAMAM dendrimers were previously shown to interact with the bilayer lipid membranes [78,79], and this therefore leaves room for the possible membrane-binding related effects on the ion channel conductance. For example, the blocker-induced PA_{63} voltage gating (the second mode of action), shown to depend on both the multivalent blocker [5] addition and the bilayer lipid composition [14], may originate from the dendrimer/membrane interaction process. At the same time, with the exception of G2-COONa (Figure S4), we did not observe any significant membrane instability upon PAMAM dendrimer addition to the membrane bathing solutions.

5. Materials and Methods

5.1. Reagents

PA_{63} was purchased from List Biological Laboratories, Inc., (Campbell, CA, USA). The following chemical reagents were used: KCl, MES, KOH, and HCl (Sigma-Aldrich, St. Louis, MO, USA), "purum" hexadecane (Fluka, Buchs, Switzerland), diphytanoylphosphatidylcholine, (DPhPC, Avanti Polar lipids,

Inc., Alabaster, AL, USA), pentane (Burdick and Jackson, Muskegon, MI, USA), and agarose (Bethesda Research Laboratory, Gaithersburg, MD, USA). MQ water was used to prepare solutions. Primary amine (generation 2) and hydroxyl (generation 2) PAMAM dendrimers, commercially available at Dendritech, Inc., (Midland, MI, USA) as w/w H$_2$O solutions, were a kind gift from Dr. Sergey Bezrukov. G3 primary amino dendrons, mixed-surface 75% OH 25% G2-NH$_2$ dendrimers, G2 carboxylate-Na terminated PAMAM dendrimers and G2 succinamic acid terminated PAMAM dendrimers were purchased from Dendritech, Inc. (Midland, MI, USA) as w/w H$_2$O solutions. Note that G2-COONa dendrimers were named G1.5 by the manufacturer because instead of only the terminal amino groups (like in the case of G2-NH$_2$ vs. G2-OH substitution), all –NH–CH$_2$–CH$_2$–NH$_2$ end groups were replaced with the COONa substituents, shortening the terminal chain of each branch by about a half. Dendritech, Inc has provided us with the analytical measurements on their PAMAM dendrimer products which are given in Figures S5 and S6. However even though mass spectrometry is often considered as the main method to characterize the presence and nature of defects in the dendrimer structure, it was demonstrated that "dendrimer purity needs to be interpreted with care and may be misleading in the sense that falsely negative results are obtained" [80].

5.2. Channel Reconstitution into Planar Lipid Bilayers

To form solvent-free planar lipid bilayers with the lipid monolayer opposition technique [81], we used a 5 mg/mL stock solution of diphytanoylphosphatidylcholine (DPhPC) in pentane. Bilayer lipid membranes were formed on a 60-μm-diameter aperture in the 15-μm-thick Teflon film that separated the electrolyte chamber into two compartments, as described in detail elsewhere [14]. The 0.1 and 1 M aqueous solutions of KCl were buffered at pH 6 (5 mM MES) at room temperature (23 $\pm$ 0.5 °C). Single channels were formed by adding 0.5 to 1 μL of 20 μg·mL^{-1} solution of PA$_{63}$ to the 1.5 mL aqueous phase in the *cis*-half of the bilayer chamber. Under this protocol, PA$_{63}$ channel insertions were always directional, as judged by channel conductance asymmetry in the applied transmembrane voltage. Multichannel experiments were performed in 0.1 M and 1 M KCl solutions, buffered at pH 6 by 5 mM MES, at 20 mV applied voltage; $\sim$1–2 μL of 1 mg·mL^{-1} stock PA$_{63}$ solution was added to the *cis*-side of the chamber. The electrical potential difference across the lipid bilayer was applied with a pair of Ag-AgCl electrodes in 2 M KCl, 1.5% agarose bridges. In all experiments, the PAMAM dendrimers were added to the *cis*-compartment of a bilayer chamber, which was the side of PA$_{63}$ addition. The *cis* compartment is believed to correspond to the endosome-facing "flower" side of the channel. Single-channel measurements were performed at -50 to $+180$ mV. Multichannel experiments were performed at 20 mV. The applied potential is defined as positive if it is higher on the side of protein addition (*cis*-side). Conductance measurements were done using an Axopatch 200B amplifier (Molecular Devices, LLC., Sunnyvale, CA, USA) in the voltage clamp mode. Signals were filtered by a low-pass 8-pole Butterworth filter (Model 9002, Frequency Devices, Inc., Haverhill, MA, USA) at 15 Hz for multichannel and 15 kHz for single channel systems and sampled with a frequency of 50 Hz and 50 kHz for multichannel and single channel experiments respectively. Amplitude, lifetime, and fluctuation analysis was performed with ClampFit 10.2 (Molecular Devices, Sunnyvale, CA, USA) and OriginPro 8.5 (OriginLab, Northampton, MA, USA) software, as well as with software developed in-house.

5.3. Reproducibility of the Experiments and Statistics

All multichannel planar lipid measurements, performed to obtain the IC_{50} data shown in Table 1, were repeated at least two times. Values are given as the means $\pm$ standard deviations. The single-channel statistical analysis with G2-NH$_2$, G2-OH, G2-SA dendrimers and G3 dendron was performed with the power spectral analysis of hundreds of dendrimer-induced current fluctuation events. At voltages >150 mV, because of the strong intrinsic and blocker-induced voltage gating, collection of a significant number of the reversible binding events proved to be difficult. Therefore the on- and off-rates were determined by averaging over lifetimes of several dozen events. Because of the wide distribution of lifetimes in the case of non-homogeneous G2 75% NH_2 25% OH dendrimer, the recordings were analyzed by averaging over the event lifetimes at all voltages. The standard deviation for t_{res} ($SD_{t_{res}}$) and k_{on} ($SD_{k_{on}}$) were then determined by averaging over 2–6 recordings taken independently from different PA_{63} channel reconstitution experiments. The standard deviations for K_D values were calculated from SDs determined for the on- and off-rates as

$$SD_{K_D} = \frac{1}{t_{res}k_{on}} \sqrt{\left(\frac{SD_{t_{res}}}{t_{res}}\right)^2 + \left(\frac{SD_{k_{on}}}{k_{on}}\right)^2}.$$

Supplementary Materials: The following are available online at www.mdpi.com/2072-6651/8/11/337/s1, Figure S1, Chemical structures of the PAMAM dendrimers used in this study. (A) Cationic PAMAM dendrimers G2-NH_2, with 16 positively charged terminal groups (left), G2-OH, with positively-charged PAMAM core and neutral OH terminal groups (right). (B) G2 PAMAM dendrimers with negatively charged succinamate (left) and carboxyl (right) terminal groups, G2-SA and G2-COONa respectively. (C) Imperfect G2 PAMAM dendrimers G2 75% OH 25% NH_2, with 12 neutral OH and 4 positively charged NH_2 terminal groups on average (left), and G3-NH_2 dendron, with a fractured more flexible structure and 16 positively charged terminal groups (right). Similar to the Figure 1B color coding, terminal primary amines are colored in red; core tertiary amines are colored in green; terminal hydroxyl groups are colored in blue. The images were created using chemical drawing software ChemDoodle 8.1.0, iChemLabs, LLC. Note that in contrast to all other dendrimers, G2 75% OH 25% NH_2 is not monodisperse and contains 75% of terminal OH groups and 25% of terminal NH_2 groups on average. Figure S2, Second mode of G2-NH_2-induced modulation of a single PA_{63} channel current. At the relatively low applied voltages (70 and 80 mV), PA_{63} mostly remains in an open state in the blocker-free solutions (left, two upper rows). Fast flickering between the open and closed states (the so-called $1/f$ noise) is mostly removed by averaging over a time interval of 100 ms. At 90 mV (left, low row), several pronounced voltage gating events are seen; $p_{bl} = 0.12$. In the presence of 0.35 µM of G2-NH_2 (right), the voltage gating of the channel is significantly increased. Multiple fast current blockages (first mode of dendrimer-induced current inhibition) are observed but they are partially filtered over a time interval of 100 ms. Figure S3, Influence of the PAMAM dendrimer terminal group chemistry on the PA_{63} channel inhibition studied on a multichannel level. (A) A typical dendrimer-induced PA_{63} inhibition curve (shown for G2-OH dendrimer). G2-OH additions are marked with the downward arrows; total bulk dendrimer concentration is indicated. The dashed line represents zero current level; (B) Typical multichannel titration curves of the PA_{63} channel inhibition by G2-NH_2, G2-OH, and G2-SA dendrimers. The dashed line represents 50% of the original current level. The recordings were taken in 0.1 M KCl solutions at pH 6 under 20 mV applied voltage. Figure S4, Effect of G2-COONa *cis*-side addition on a single PA_{63} channel. Both in blocker-free solution and in presence of 3 mM G2-COONa, PA_{63} mostly remains in an open state. The $1/f$ events are to a large extent removed by averaging over a time interval of 10 ms. G2-COONa addition causes lipid bilayer instability (upward events) and, eventually, breakage. Recordings were taken in 1 M KCl solutions at pH 6 and 100 mV applied voltage. Figure S5, MALDI-TOF mass spectra of G2-NH_2 (A); G2-OH (B); G2-SA (C); G2-COONa (D) dendrimers and G3-NH_2 dendron (E). The data were provided by Dendritech, Inc. (Midland, MI, USA). Figure S6, Characterization of generation 2 PAMAM dendrimers and generation 3 PAMAM-NH_2 dendron by 13 C NMR. The data were provided by Dendritech, Inc. (Midland, MI, USA), (A) 13 C NMR of G2-NH_2. (B) 13 C NMR of G2-OH; (C) 13 C NMR of G2-SA; (D) 13 C NMR of G2-COONa; (E) 13 C NMR G3-NH_2 dendron.

Acknowledgments: The project was financially supported by National Institute of Allergy and Infectious Diseases of the National Institutes of Health under award number 1R15AI099897-01A1 and by The Catholic University startup funds (to Ekaterina Nestorovich). Many thanks to Sergey M. Bezrukov and Edward L. Mertz (both NICHD, NIH) for fruitful discussion.

Author Contributions: Goli Yamini and Ekaterina M. Nestorovich designed and performed experiments, analyzed the data, and wrote the manuscript. Nnanya Kalu performed experiments and contributed to the manuscript proofreading.

Conflicts of Interest: The authors declare no conflict of interest.

References

1. Yamini, G.; Nestorovich, E.M. Multivalent inhibitors of channel-forming bacterial toxins. *Curr. Top. Microbiol. Immunol.* **2016**. [CrossRef]

2. Karginov, V.A.; Nestorovich, E.M.; Moayeri, M.; Leppla, S.H.; Bezrukov, S.M. Blocking anthrax lethal toxin at the protective antigen channel by using structure-inspired drug design. *Proc. Natl. Acad. Sci. USA* **2005**, *102*, 15075–15080. [CrossRef] [PubMed]

3. Nestorovich, E.M.; Karginov, V.A.; Popoff, M.R.; Bezrukov, S.M.; Barth, H. Tailored ß-cyclodextrin blocks the translocation pores of binary exotoxins from *C. botulinum* and *C. perfringens* and protects cells from intoxication. *PLoS ONE* **2011**, *6*. [CrossRef] [PubMed]

4. Roeder, M.; Nestorovich, E.M.; Karginov, V.A.; Schwan, C.; Aktories, K.; Barth, H. Tailored cyclodextrin pore blocker protects mammalian cells from clostridium difficile binary toxin CDT. *Toxins (Basel)* **2014**, *6*, 2097–2114. [CrossRef] [PubMed]
5. Forstner, P.; Bayer, F.; Kalu, N.; Felsen, S.; Fortsch, C.; Aloufi, A.; Ng, D.Y.; Weil, T.; Nestorovich, E.M.; Barth, H. Cationic PAMAM dendrimers as pore-blocking binary toxin inhibitors. *Biomacromolecules* **2014**, *15*, 2461–2474. [CrossRef] [PubMed]
6. Nestorovich, E.M.; Bezrukov, S.M. Obstructing toxin pathways by targeted pore blockage. *Chem. Rev.* **2012**, *112*, 6388–6430. [CrossRef] [PubMed]
7. Krantz, B.A.; Melnyk, R.A.; Zhang, S.; Juris, S.J.; Lacy, D.B.; Wu, Z.; Finkelstein, A.; Collier, R.J. A phenylalanine clamp catalyzes protein translocation through the anthrax toxin pore. *Science* **2005**, *309*, 777–781. [CrossRef] [PubMed]
8. Bachmeyer, C.; Orlik, F.; Barth, H.; Aktories, K.; Benz, R. Mechanism of C2-toxin inhibition by fluphenazine and related compounds: Investigation of their binding kinetics to the C2II-channel using the current noise analysis. *J. Mol. Biol.* **2003**, *333*, 527–540. [CrossRef] [PubMed]
9. Orlik, F.; Schiffler, B.; Benz, R. Anthrax toxin protective antigen: Inhibition of channel function by chloroquine and related compounds and study of binding kinetics using the current noise analysis. *Biophys. J.* **2005**, *88*, 1715–1724. [CrossRef] [PubMed]
10. Beitzinger, C.; Bronnhuber, A.; Duscha, K.; Riedl, Z.; Huber-Lang, M.; Benz, R.; Hajós, G.; Barth, H. Designed azolopyridinium salts block protective antigen pores in vitro and protect cells from anthrax toxin. *PLoS ONE* **2013**, *8*. [CrossRef] [PubMed]
11. Kronhardt, A.; Beitzinger, C.; Barth, H.; Benz, R. Chloroquine Analog Interaction with C2- and Iota-Toxin in Vitro and in Living Cells. *Toxins (Basel)* **2016**, *8*. [CrossRef] [PubMed]
12. Crini, G. Review: A history of cyclodextrins. *Chem. Rev.* **2014**, *114*, 10940–10975. [CrossRef] [PubMed]
13. Wu, L.P.; Ficker, M.; Christensen, J.B.; Trohopoulos, P.N.; Moghimi, S.M. Dendrimers in medicine: Therapeutic concepts and pharmaceutical challenges. *Bioconjug. Chem.* **2015**, *26*, 1198–1211. [CrossRef] [PubMed]
14. Nestorovich, E.M.; Karginov, V.A.; Berezhkovskii, A.M.; Bezrukov, S.M. Blockage of anthrax PA63 pore by a multicharged high-affinity toxin inhibitor. *Biophys. J.* **2010**, *99*, 134–143. [CrossRef] [PubMed]
15. Bezrukov, S.M.; Liu, X.; Karginov, V.A.; Wein, A.N.; Leppla, S.H.; Popoff, M.R.; Barth, H.; Nestorovich, E.M. Interactions of high-affinity cationic blockers with the translocation pores of *B. anthracis*, *C. botulinum*, and *C. perfringens* binary toxins. *Biophys. J.* **2012**, *103*, 1208–1217. [CrossRef] [PubMed]
16. Lee, C.C.; MacKay, J.A.; Frechet, J.M.; Szoka, F.C. Designing dendrimers for biological applications. *Nat. Biotechnol.* **2005**, *23*, 1517–1526. [CrossRef] [PubMed]
17. Svenson, S.; Tomalia, D.A. Dendrimers in biomedical applications-reflections on the field. *Adv. Drug Deliv. Rev.* **2005**, *57*, 2106–2129. [CrossRef] [PubMed]
18. Duncan, R.; Izzo, L. Dendrimer biocompatibility and toxicity. *Adv. Drug Deliv. Rev.* **2005**, *57*, 2215–2237. [CrossRef] [PubMed]
19. Hong, S.; Bielinska, A.U.; Mecke, A.; Keszler, B.; Beals, J.L.; Shi, X.; Balogh, L.; Orr, B.G.; Baker, J.R., Jr.; Banaszak Holl, M.M. Interaction of poly(amidoamine) dendrimers with supported lipid bilayers and cells: Hole formation and the relation to transport. *Bioconjug. Chem.* **2004**, *15*, 774–782. [CrossRef] [PubMed]
20. Tang, M.X.; Redemann, C.T.; Szoka, F.C., Jr. In vitro gene delivery by degraded polyamidoamine dendrimers. *Bioconjug. Chem.* **1996**, *7*, 703–714. [CrossRef] [PubMed]
21. Nestorovich, E.M.; Bezrukov, S.M. Designing inhibitors of anthrax toxin. *Expert Opin. Drug Discov.* **2014**, *9*, 299–318. [CrossRef] [PubMed]
22. Duesbery, N.S.; Webb, C.P.; Leppla, S.H.; Gordon, V.M.; Klimpel, K.R.; Copeland, T.D.; Ahn, N.G.; Oskarsson, M.K.; Fukasawa, K.; Paull, K.D.; et al. Proteolytic inactivation of MAP-kinase-kinase by anthrax lethal factor. *Science* **1998**, *280*, 734–737. [CrossRef] [PubMed]
23. Vitale, G.; Bernardi, L.; Napolitani, G.; Mock, M.; Montecucco, C. Susceptibility of mitogen-activated protein kinase kinase family members to proteolysis by anthrax lethal factor. *Biochem. J.* **2000**, *352*, 739–745. [CrossRef] [PubMed]
24. Levinsohn, J.L.; Newman, Z.L.; Hellmich, K.A.; Fattah, R.; Getz, M.A.; Liu, S.; Sastalla, I.; Leppla, S.H.; Moayeri, M. Anthrax lethal factor cleavage of Nlrp1 is required for activation of the inflammasome. *PLoS Pathog.* **2012**, *8*. [CrossRef] [PubMed]

25. Leppla, S.H. Anthrax toxin edema factor: A bacterial adenylate cyclase that increases cyclic AMP concentrations of eukaryotic cells. *Proc. Natl. Acad. Sci. USA* **1982**, *79*, 3162–3166. [CrossRef] [PubMed]

26. Leppla, S.H. *Bacillus anthracis* calmodulin-dependent adenylate cyclase: Chemical and enzymatic properties and interactions with eucaryotic cells. *Adv. Cycl. Nucleotide Protein Phosphorylation Res.* **1984**, *17*, 189–198.

27. Petosa, C.; Collier, R.J.; Klimpel, K.R.; Leppla, S.H.; Liddington, R.C. Crystal structure of the anthrax toxin protective antigen. *Nature* **1997**, *385*, 833–838. [CrossRef] [PubMed]

28. Kintzer, A.F.; Thoren, K.L.; Sterling, H.J.; Dong, K.C.; Feld, G.K.; Tang, I.I.; Williams, E.R.; Berger, J.M.; Krantz, B.A. The protective antigen component of anthrax toxin forms functional octameric complexes. *J. Mol. Biol.* **2009**, *392*, 614–629. [CrossRef] [PubMed]

29. Kintzer, A.F.; Sterling, H.J.; Tang, I.I.; Williams, E.R.; Krantz, B.A. Anthrax toxin receptor drives protective antigen oligomerization and stabilizes the heptameric and octameric oligomer by a similar mechanism. *PLoS ONE* **2010**, *5*. [CrossRef] [PubMed]

30. Mogridge, J.; Cunningham, K.; Collier, R.J. Stoichiometry of anthrax toxin complexes. *Biochemistry* **2002**, *41*, 1079–1082. [CrossRef] [PubMed]

31. Pilpa, R.M.; Bayrhuber, M.; Marlett, J.M.; Riek, R.; Young, J.A. A receptor-based switch that regulates anthrax toxin pore formation. *PLoS Pathog.* **2011**, *7*. [CrossRef] [PubMed]

32. Abrami, L.; Brandi, L.; Moayeri, M.; Brown, M.J.; Krantz, B.A.; Leppla, S.H.; van der Goot, F.G. Hijacking multivesicular bodies enables long-term and exosome-mediated long-distance action of anthrax toxin. *Cell. Rep.* **2013**, *5*, 986–996. [CrossRef] [PubMed]

33. Blaustein, R.O.; Koehler, T.M.; Collier, R.J.; Finkelstein, A. Anthrax toxin: Channel-forming activity of protective antigen in planar phospholipid bilayers. *Proc. Natl. Acad. Sci. USA* **1989**, *86*, 2209–2213. [CrossRef] [PubMed]

34. Jiang, J.; Pentelute, B.L.; Collier, R.J.; Zhou, Z.H. Atomic structure of anthrax protective antigen pore elucidates toxin translocation. *Nature* **2015**, *521*, 545–549. [CrossRef] [PubMed]

35. Zhang, S.; Udho, E.; Wu, Z.; Collier, R.J.; Finkelstein, A. Protein translocation through anthrax toxin channels formed in planar lipid bilayers. *Biophys. J.* **2004**, *87*, 3842–3849. [CrossRef] [PubMed]

36. Zhang, S.; Finkelstein, A.; Collier, R.J. Evidence that translocation of anthrax toxin's lethal factor is initiated by entry of its N terminus into the protective antigen channel. *Proc. Natl. Acad. Sci. USA* **2004**, *101*, 16756–16761. [CrossRef] [PubMed]

37. Nablo, B.J.; Panchal, R.G.; Bavari, S.; Nguyen, T.L.; Gussio, R.; Ribot, W.; Friedlander, A.; Chabot, D.; Reiner, J.E.; Robertson, J.W.; et al. Anthrax toxin-induced rupture of artificial lipid bilayer membranes. *J. Chem. Phys.* **2013**, *139*. [CrossRef] [PubMed]

38. Rappaport, S.M.; Teijido, O.; Hoogerheide, D.P.; Rostovtseva, T.K.; Berezhkovskii, A.M.; Bezrukov, S.M. Conductance hysteresis in the voltage-dependent anion channel. *Eur. Biophys. J.* **2015**, *44*, 465–472. [CrossRef] [PubMed]

39. Sen, K.; Hellman, J.; Nikaido, H. Porin channels in intact cells of Escherichia coli are not affected by Donnan potentials across the outer membrane. *J. Biol. Chem.* **1988**, *263*, 1182–1187. [PubMed]

40. Robertson, K.M.; Tieleman, D.P. Molecular basis of voltage gating of OmpF porin. *Biochem. Cell Biol.* **2002**, *80*, 517–523. [CrossRef] [PubMed]

41. Delcour, A.H. *Electrophysiology of Unconventional Channels and Pores*; Springer: Cham, Switzerland, 2015.

42. Bainbridge, G.; Gokce, I.; Lakey, J.H. Voltage gating is a fundamental feature of porin and toxin beta-barrel membrane channels. *FEBS Lett.* **1998**, *431*, 305–308. [CrossRef]

43. Teijido, O.; Rappaport, S.M.; Chamberlin, A.; Noskov, S.Y.; Aguilella, V.M.; Rostovtseva, T.K.; Bezrukov, S.M. Acidification asymmetrically affects voltage-dependent anion channel implicating the involvement of salt bridges. *J. Biol. Chem.* **2014**, *289*, 23670–23682. [CrossRef] [PubMed]

44. Rostovtseva, T.K.; Tan, W.; Colombini, M. On the role of VDAC in apoptosis: Fact and fiction. *J. Bioenerg. Biomembr.* **2005**, *37*, 129–142. [CrossRef] [PubMed]

45. Kullman, L.; Winterhalter, M.; Bezrukov, S.M. Transport of maltodextrins through maltoporin: A single-channel study. *Biophys. J.* **2002**, *82*, 803–812. [CrossRef]

46. Blaustein, R.O.; Finkelstein, A. Diffusion limitation in the block by symmetric tetraalkylammonium ions of anthrax toxin channels in planar phospholipid bilayer membranes. *J. Gen. Physiol.* **1990**, *96*, 943–957. [CrossRef] [PubMed]

47. Blaustein, R.O.; Finkelstein, A. Voltage-dependent block of anthrax toxin channels in planar phospholipid bilayer membranes by symmetric tetraalkylammonium ions. Effects on macroscopic conductance. *J. Gen. Physiol.* **1990**, *96*, 905–919. [CrossRef] [PubMed]

48. Karginov, V.A.; Nestorovich, E.M.; Yohannes, A.; Robinson, T.M.; Fahmi, N.E.; Schmidtmann, F.; Hecht, S.M.; Bezrukov, S.M. Search for cyclodextrin-based inhibitors of anthrax toxins: Synthesis, structural features, and relative activities. *Antimicrob. Agents Chemother.* **2006**, *50*, 3740–3753. [CrossRef] [PubMed]

49. Anderson, D.S.; Blaustein, R.O. Preventing voltage-dependent gating of anthrax toxin channels using engineered disulfides. *J. Gen. Physiol.* **2008**, *132*, 351–360. [CrossRef] [PubMed]

50. Moayeri, M.; Robinson, T.M.; Leppla, S.H.; Karginov, V.A. In vivo efficacy of beta-cyclodextrin derivatives against anthrax lethal toxin. *Antimicrob. Agents Chemother.* **2008**, *52*, 2239–2241. [CrossRef] [PubMed]

51. French, R.J.; Shoukimas, J.J. An ion's view of the potassium channel. The structure of the permeation pathway as sensed by a variety of blocking ions. *J. Gen. Physiol.* **1985**, *85*, 669–698. [CrossRef] [PubMed]

52. Gurnev, P.A.; Queralt-Martin, M.; Aguilella, V.M.; Rostovtseva, T.K.; Bezrukov, S.M. Probing tubulin-blocked state of VDAC by varying membrane surface charge. *Biophys. J.* **2012**, *102*, 2070–2076. [CrossRef] [PubMed]

53. Gurnev, P.A.; Yap, T.L.; Pfefferkorn, C.M.; Rostovtseva, T.K.; Berezhkovskii, A.M.; Lee, J.C.; Parsegian, V.A.; Bezrukov, S.M. Alpha-synuclein lipid-dependent membrane binding and translocation through the alpha-hemolysin channel. *Biophys. J.* **2014**, *106*, 556–565. [CrossRef] [PubMed]

54. Tomalia, D.A.; Frechet, J.M.J. Discovery of dendrimers and dendritic polymers: A brief historical perspective. *J. Polym. Sci. A Polym. Chem.* **2002**, *40*, 2719–2728. [CrossRef]

55. Tomalia, D.A.; Baker, H.; Dewald, J.; Hall, M.; Kallos, G.; Martin, S.; Roeck, J.; Ryder, J.; Smith, P. A new class of polymers: Starburst-dendritic macromolecules. *Polym. J.* **1985**, *17*, 117–132. [CrossRef]

56. Ballauff, M.; Likos, C.N. Dendrimers in solution: Insight from theory and simulation. *Angew. Chem. Int. Ed. Engl.* **2004**, *43*, 2998–3020. [CrossRef] [PubMed]

57. Huissmann, S.; Likos, C.N.; Blaak, R. Conformations of high-generation dendritic polyelectrolytes. *J. Mater. Chem.* **2010**, *20*, 10486–10494. [CrossRef]

58. Maiti, P.K.; Lin, S.T.; Cagin, T.; Goddard, W.A. Effect of Solvent and pH on the Structure of PAMAM Dendrimers. *Macromolecules* **2005**, *38*, 979–991. [CrossRef]

59. Huissmann, S.; Wynveen, A.; Likos, C.N.; Blaak, R. The effects of pH, salt and bond stiffness on charged dendrimers. *J. Phys. Condens Matter* **2010**, *22*, 232101. [CrossRef] [PubMed]

60. Liu, Y.; Bryantsev, V.S.; Diallo, M.S.; Goddard, W.A., 3rd. PAMAM dendrimers undergo pH responsive conformational changes without swelling. *J. Am. Chem. Soc.* **2009**, *131*, 2798–2799. [CrossRef] [PubMed]

61. Garcia-Fernandez, E.; Paulo, P.M. Deswelling and electrolyte dissipation in free diffusion of charged PAMAM dendrimers. *J. Phys. Chem. Lett.* **2014**, *5*, 1472–1478. [CrossRef] [PubMed]

62. Maiti, P.K.; Bagchi, B. Diffusion of flexible, charged, nanoscopic molecules in solution: Size and pH dependence for PAMAM dendrimer. *J. Chem. Phys.* **2009**, *131*, 214901. [CrossRef] [PubMed]

63. Maiti, P.K.; Messina, R. Counterion Distribution and ζ-Potential in PAMAM Dendrimer. *Macromolecules* **2008**, *41*, 5002–5006. [CrossRef]

64. Mecke, A.; Lee, I.; Baker, J.R., Jr.; Holl, M.M.; Orr, B.G. Deformability of poly(amidoamine) dendrimers. *Eur. Phys. J. E. Soft Matter* **2004**, *14*, 7–16. [CrossRef] [PubMed]

65. Liu, Y.; Porcar, L.; Hong, K.; Shew, C.Y.; Li, X.; Liu, E.; et al. Effect of counterion valence on the pH responsiveness of polyamidoamine dendrimer structure. *J. Chem. Phys.* **2010**, *132*. [CrossRef] [PubMed]

66. Porcar, L.; Liu, Y.; Verduzco, R.; Hong, K.; Butler, P.D.; Magid, L.J.; Smith, G.S.; Chen, W. Structural investigation of PAMAM dendrimers in aqueous solutions using small-angle neutron scattering: Effect of generation. *J. Phys. Chem. B* **2008**, *112*, 14772–14778. [CrossRef] [PubMed]

67. Wu, B.; Kerkeni, B.; Egami, T.; Do, C.; Liu, Y.; Wang, Y.; Porcar, L.; Hong, K.; Smith, S.C.; Liu, E.L.; et al. Structured water in polyelectrolyte dendrimers: Understanding small angle neutron scattering results through atomistic simulation. *J. Chem. Phys.* **2012**, *136*, 144901. [CrossRef] [PubMed]

68. Cakara, D.; Kleimann, J.; Borkovec, M. Microscopic protonation equilibria of poly(amidoamine) dendrimers from macroscopic titrations. *Macromolecules* **2003**, *36*, 4201–4207. [CrossRef]

69. Huang, Q.R.; Dubin, P.L.; Moorefield, C.N.; Newkome, G.R. Counterion binding on charged spheres: Effect of pH and ionic strength on the mobility of carboxyl-terminated dendrimers. *J. Phys. Chem. B* **2000**, *104*, 898–904. [CrossRef]

70. Dobrovolskaia, M.A.; Patri, A.K.; Simak, J.; Hall, J.B.; Semberova, J.; de Paoli Lacerda, S.H.; McNeil, S.E. Nanoparticle size and surface charge determine effects of PAMAM dendrimers on human platelets in vitro. *Mol. Pharm.* **2012**, *9*, 382–393. [CrossRef] [PubMed]

71. Böhme, U.; Klenge, A.; Hänel, B.; Scheler, U. Counterion condensation and effective charge of PAMAM dendrimers. *Polymers* **2011**, *3*, 812–819. [CrossRef]

72. Bustamante, J.O.; Michelette, E.R.; Geibel, J.P.; Hanover, J.A.; McDonnell, T.J.; Dean, D.A. Dendrimer-assisted patch-clamp sizing of nuclear pores. *Pflugers Arch.* **2000**, *439*, 829–837. [CrossRef] [PubMed]

73. Martin, H.; Kinns, H.; Mitchell, N.; Astier, Y.; Madathil, R.; Howorka, S. Nanoscale protein pores modified with PAMAM dendrimers. *J. Am. Chem. Soc.* **2007**, *129*, 9640–9649. [CrossRef] [PubMed]

74. Kong, L.; Harrington, L.; Li, Q.; Cheley, S.; Davis, B.G.; Bayley, H. Single-molecule interrogation of a bacterial sugar transporter allows the discovery of an extracellular inhibitor. *Nat. Chem.* **2013**, *5*, 651–659. [CrossRef] [PubMed]

75. Ficici, E.; Andricioaei, I.; Howorka, S. Dendrimers in nanoscale confinement: The interplay between conformational change and nanopore entrance. *Nano Lett.* **2015**, *15*, 4822–4828. [CrossRef] [PubMed]

76. Mandal, T.; Kanchi, S.; Ayappa, K.G.; Maiti, P.K. pH controlled gating of toxic protein pores by dendrimers. *Nanoscale* **2016**, *8*, 13045–13058. [CrossRef] [PubMed]

77. Rostovtseva, T.K.; Gurnev, P.A.; Chen, M.Y.; Bezrukov, S.M. Membrane lipid composition regulates tubulin interaction with mitochondrial voltage-dependent anion channel. *J. Biol. Chem.* **2012**, *287*, 29589–29598. [CrossRef] [PubMed]

78. Klajnert, B.; Epand, R.M. PAMAM dendrimers and model membranes: Differential scanning calorimetry studies. *Int. J. Pharm.* **2005**, *305*, 154–166. [CrossRef] [PubMed]

79. Mecke, A.; Uppuluri, S.; Sassanella, T.M.; Lee, D.K.; Ramamoorthy, A.; Baker, J.R., Jr.; Orr, B.G.; Banaszak Holl, M.M. Direct observation of lipid bilayer disruption by poly(amidoamine) dendrimers. *Chem. Phys. Lipids* **2004**, *132*, 3–14. [CrossRef] [PubMed]

80. Baytekin, B.; Werner, N.; Luppertz, F.; Engeser, M.; Brüggemann, J.; Bitter, S.; Henkel, R.; Felder, T.; Schalley, C.A. How useful is mass spectrometry for the characterization of dendrimers? "Fake defects" in the ESI and MALDI mass spectra of dendritic compounds. *Int. J. Mass Spectrom.* **2006**, *249–250*, 138–148. [CrossRef]

81. Montal, M.; Mueller, P. Formation of bimolecular membranes from lipid monolayers and a study of their electrical properties. *Proc. Natl. Acad. Sci. USA* **1972**, *69*, 3561–3566. [CrossRef] [PubMed]

Article

Chloroquine Analog Interaction with C2- and Iota-Toxin in Vitro and in Living Cells

Angelika Kronhardt [1,†], Christoph Beitzinger [1,†], Holger Barth [2] and Roland Benz [3,*]

1   Rudolf Virchow Center, Research Center for Experimental Biomedicine, University of Würzburg,
    Versbacher Straße 9, 97078 Würzburg, Germany; ange.kronhardt@web.de (A.K.);
    christoph.beitzinger@web.de (C.B.)
2   Institute of Pharmacology and Toxicology, University of Ulm Medical Center, Albert-Einstein-Allee 11,
    89081 Ulm, Germany; holger.barth@uni-ulm.de
3   Department of Life Sciences and Chemistry, Jacobs-University Bremen, Campus-Ring 1,
    28759 Bremen, Germany
*   Correspondence: r.benz@jacobs-university.de; Tel.: +49-421-200-3151
†   These authors contributed equally to this work.

Academic Editor: Michel R. Popoff
Received: 6 June 2016; Accepted: 28 July 2016; Published: 10 August 2016

Abstract: C2-toxin from *Clostridium botulinum* and Iota-toxin from *Clostridium perfringens* belong both to the binary A-B-type of toxins consisting of two separately secreted components, an enzymatic subunit A and a binding component B that facilitates the entry of the corresponding enzymatic subunit into the target cells. The enzymatic subunits are in both cases actin ADP-ribosyltransferases that modify R177 of globular actin finally leading to cell death. Following their binding to host cells' receptors and internalization, the two binding components form heptameric channels in endosomal membranes which mediate the translocation of the enzymatic components Iota a and C2I from endosomes into the cytosol of the target cells. The binding components form ion-permeable channels in artificial and biological membranes. Chloroquine and related 4-aminoquinolines were able to block channel formation in vitro and intoxication of living cells. In this study, we extended our previous work to the use of different chloroquine analogs and demonstrate that positively charged aminoquinolinium salts are able to block channels formed in lipid bilayer membranes by the binding components of C2- and Iota-toxin. Similarly, these molecules protect cultured mammalian cells from intoxication with C2- and Iota-toxin. The aminoquinolinium salts did presumably not interfere with actin ADP-ribosylation or receptor binding but blocked the pores formed by C2IIa and Iota b in living cells and in vitro. The blocking efficiency of pores formed by Iota b and C2IIa by the chloroquine analogs showed interesting differences indicating structural variations between the types of protein-conducting nanochannels formed by Iota b and C2IIa.

Keywords: C2-toxin; iota-toxin; binding components; chloroquine; black lipid bilayer; aminoquinolinium salts

1. Introduction

Binary A-B type protein toxins are potent virulence factors of certain gram-positive bacteria (for reviews see refs [1–3]). The most prominent example of this type of toxins is the anthrax toxin produced by *Bacillus anthracis*, which is also known as a possible biological weapon [4–6]. Other prominent examples are C2-toxin of *Clostridium botulinum* and Iota-toxin of *Clostridium perfringens*. Both toxins consist of two distinct components that are secreted separately into the extracellular media: an enzymatically active component A—which acts as an actin-specific ADP-ribosyltransferase—and a separate component

B, which is the binding/translocation subunit needed for binding of the toxins to target cells and responsible for translocation of the enzymatic subunits into the cytosol of target cells [4,7–12].

After the proteolytic activation the B components C2II of *Clostridium botulinum* and also Iota b of *Clostridium perfringens* form ring-shaped heptamers similar to the B component of the anthrax toxin PA [11,13–16]. These heptamers (C2IIa, Iota b) are the biologically active species of the B components and mediate two different functions during cellular uptake of the toxins: First, they bind to their receptors on the surface of target cells and form complexes with their A components. These complexes are subsequently taken up into cells via receptor-mediated endocytosis and thereby reach early endosomal vesicles. The acidic conditions in such endosomes trigger a conformational change of the compound B heptamers, which insert into endosomal membranes to form trans-membrane pores. These pores serve as translocation channels for the subsequent transport of the unfolded A components of these toxins from the endosomal lumen into the host cell cytosol. Treatment of cells with bafilomycin (Baf) A1, a compound that prevents acidification of the endosomes, inhibits pore-formation by the B components, and therefore the translocation of the A components across endosomal membranes into the cytosol and thus protects cells from intoxication with these toxins [1,17–20]. Such a translocation mechanism is common to other binary toxins, including anthrax toxin from *Bacillus anthracis* [1,21].

The enzymatic components develop their activity in the cytosol of the target cells where they ADP-ribosylate monomeric G-actin at position arginine 177 with NAD as co-substrate leading to actin depolymerization, cell rounding, and eventually cell death [1,22–26]. Similarly, other members of the family of binary toxins act also as ADP-ribosylating toxins. These are CDT (*Clostridium difficile* binary toxin) of *Clostridium difficile* [27–29], *Clostridium spiroforme* toxin [30], and the vegetative insecticidal proteins (VIPs) of *Bacillus cere*us and *Bacillus thuringiensis* [31,32].

The inhibition of channel function by binding components and intoxication of target cells by compounds that bind to the binding components is of considerable interest because of the possible use of A-B type of toxins as biological weapons. Possible candidates are tailored azolopyridinium salts and tailored cyclic dextrines [33–36]. In previous studies, we have demonstrated that low concentrations of chloroquine were able to inhibit intoxication of target cells by C2-toxin in cell-based assays and pore-formation by C2IIa in lipid bilayer membranes [37,38]. Similarly, blockage of iota b channels by chloroquine was also observed in reconstitution experiments with lipid bilayers but at much higher concentrations than those needed in experiments with C2IIa [39,40]. The binding site for chloroquine and related compounds in the channel formed by C2IIa was identified in the vestibule on the cis-side of the mushroom-sized heptamers that corresponds to the cell surface exposed side [41]. It is presumably the same binding site that also interacts also with the positively charged N-terminus of the enzymatic subunits C2I and Iota b and directs them to the channel lumen and further on into the cytosol of the target cells [1,3,40]. This means that binding is the prerequisite for transport. Site-directed mutagenesis of E399, D426, and F428 (corresponding to the Φ–clamp in PA [42,43]) in C2IIa has clearly demonstrated that these three amino acids are elements of the binding site within the vestibule of the channel formed by C2II [41]. These amino acids are also present in the primary sequence of Iota b in similar positions (D386, D413, and F415) and there exists no doubt that they are also involved in the binding site of the heptameric Iota b channel [40]. Besides these amino acids that are directly involved in binding of Iota a and chloroquine the sequence of Iota b also contains several threonines (T292 and T320) that are probably involved in the structure and stability of the pore-forming heptamers of Iota b. Their replacement by other amino acids leads to misfolded Iota b channels that have completely different properties than the ones formed by wildtype Iota b channels [40].

In this study, we investigated the binding of different chloroquine analogs to the channels formed by the binding components C2IIa and Iota b. The interaction between the protein-conducting nanopores and the different ligands was performed by titration experiments with artificial membranes containing C2IIa and Iota b channels. This type of investigation using the dose-dependent decrease of membrane conductance allowed a rapid and meaningful investigation of the affinity of the different chloroquine analogs to the binding site inside the vestibule of the heptameric channels. Similarly, we investigated the effect of the chloroquine analog with the highest affinity for binding to C2IIa and Iota b on the pH-dependent trans-membrane transport of the A components of C2- and Iota-toxin through the trans-membrane pores formed by the B components of these toxins in living cells. The results suggested indeed that this compound blocked the trans-membrane transport of these binary toxins with much higher efficiency than chloroquine.

2. Results

2.1. Binding of Different Aminoquinolinium Salts to the Channels Formed by the Binding Components C2IIa and Iota b

The channels formed by the binding components C2IIa and Iota b are fully oriented in artificial and presumably also in biological membranes when they are added to only one side of an artificial and biological membrane [37,39]. Most of the water-soluble part of the mushroom-sized heptamer is localized on the *cis*-side of the membrane (the side of addition of the binding components). Only a few amino acids at the end of the beta-barrel cylinder of 14 beta-strands are directed to the trans-side of the membrane. The structure is similar to that of α-toxin of *Staphylococcus aureus* and the recently elucidated 3D-structure of the membrane-spanning form of the PA_{63}-channel, which forms also a heptamer with some sort of vestibule on the *cis*-side [16,44]. In previous studies we demonstrated that reconstituted C2IIa channels as well as Iota b channels can be blocked in lipid bilayer membranes by the addition of 4-aminoquinolines [38,39,45] and identified the binding site for chloroquine to C2IIa channels on the *cis* side of the C2IIa heptamer within the vestibule of the channels [41]. The binding affinity strongly depends on negatively charged amino acids and also on the Φ-clamp within the vestibule of the C2IIa and Iota b channels. The stability constant K for ligand binding to the C2IIa and the Iota b channels was calculated by multi-channel titration experiments [37–40]. Similar experiments were performed here with the chloroquine analogs C 23, C 164, C 268, and C 280. Figure 1 shows an experiment of this type. Activated Iota b was added in a concentration of about 20 ng/mL to the *cis*-side (the side of the applied potential) of a black lipid bilayer membrane while stirring. The reconstitution of Iota b channels led to a substantial increase of membrane conductivity by several orders of magnitude caused by insertion of Iota b channels in the membrane monitored by a strip chart recorder. After about 30 min to several hours, when the membrane conductance was virtually stationary, the titration experiments started. Small amounts of concentrated solution of C 164 were added after about two hours after the start of the experiment to the aqueous phase on the *cis*-side of the membrane while stirring to allow equilibration. Subsequently, the Iota b channels were blocked and the dose-dependent decrease of conductance was measured as a function of time (see Figure 1).

The analysis of the data of Figure 1 indicated that the Iota b channels were not fully blocked by the addition of C 164 at a concentration of 4.03 mM. This was caused by the problem to reach sufficiently high concentrations of the chloroquine analog, which are limited by its solubility in aqueous salt solution. The fit of the titration data shown in Figure 2A suggest in principle that the Iota b channel was only blocked by about 50%. However, when the concentration of C 164 was extrapolated to higher ones (see Figure 2B), then it was clear that compound C 164 was also able to almost fully block the Iota b channel. The stability constant of binding of C 164 to the Iota b channels was about (348 ± 48) 1/M and the channel block was at maximum 92% $\pm$ 7%. This was a very low stability constant for binding of a channel blocker to one of the binding component channels. However, the data of binding of all chloroquine analogs and chloroquine itself demonstrates that the iota b channel is not a good target for binding of chloroquine and the different chloroquine analogs (aminoquinolinium salts) (see Table 1). Only chloroquine itself and C 280 had a reasonably high affinity to the Iota b channel. Chloroquine analog C 280 that has a permanent positive charge and a bulky side chain was used for the study of the inhibition of cell cytotoxicity by Iota-toxin (see Figure 3).

The affinity of chloroquine and the chloroquine analogs to the channels formed by C2IIa was definitely more substantial as the summary of stability constants and half saturation constants of Table 1 clearly demonstrates. Chloroquine and all aminoquinolinium salts used in this study have a higher affinity to the C2IIa channel than to that formed by Iota b. With the aminoquinolinium salts, the affinity to the C2IIa channels increased in the series C 23, C 164, C 268, and C 280 by factor of more than 4000. C 268 had already a binding constant to C2IIa channels that was about twofold higher than chloroquine. The highest stability constant for binding to the C2IIa channel was C 280, which had a half saturation constant of 0.16 μM. This low K_S value for binding of C 280 to C2IIa is about a factor of 60 smaller than that of chloroquine binding which is remarkable and suggested indeed that C 280 could serve as an inhibitor of intoxication by C2-toxin.

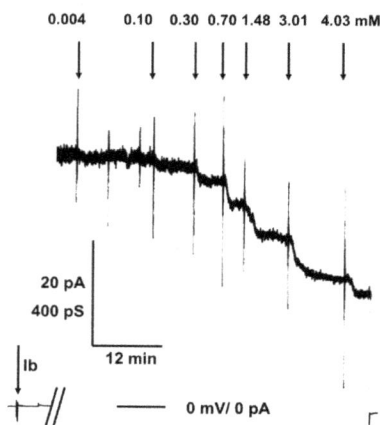

Figure 1. Titration experiment of Iota b-induced membrane conductance with C 164. The membrane was formed from diphytanoyl phosphatidylcholine/n-decane. The aqueous phase contained 20 ng/mL Iota b (Ib) protein (added to the *cis*-side of the membrane at the time of the left side arrow), 150 mM KCl, 10 mM MES, pH 6.0. The temperature was constantly 20 °C and the applied voltage was 50 mV. The two bars at the base line indicate the time interval of about 1h and 50 min between the addition of Ib and the start of the titration experiment. The membrane contained about 70 Iota b-channels (single channel conductance G = 15 pS) when C 164 was added at the indicated concentrations to the aqueous phase. The bottom line represents zero level of conductance (at begin of the experiment) or also zero level of current when voltage was switched off (at the end of titrations). Note that the high noise of the current recording during the titration experiment was caused by stirring in the membrane cell to allow rapid equilibration of C 164 in the aqueous phase.

Table 1. Stability constants K for the block of binding components channels formed by C2II and Iota b by chloroquine and related aminoquinolinium salts in lipid bilayer membranes [a]. * The results of similar titration experiments performed with C2IIa channels and chloroquine are given for comparison.

Chloroquine Analog	$K/10^3$ M^{-1}	K_S/μM	$K/10^3$ M^{-1}	K_S/μM
-	C2II		Iota b	
Chloroquine	110 *	9.1 *	7.1 ± 1.7	140
C 23	1.5 ± 0.4	710	0.82 ± 0.21	1200
C 164	18.5 ± 2.5	54	0.39 ± 0.13	2400
C 268	198 ± 15	5.1	2.5 ± 0.4	400
C 280	6200 ± 40	0.16	12.5 ± 2.1	80

[a] The data represent means ± SD of at least three individual titration experiments. The membranes were formed from diphytanoyl phosphatidylcholine/n-decane. The aqueous phase contained 150 mM KCl, 10 mM MES-KOH, pH 6, and about 10 ng/mL activated C2II or about 20 ng/mL Iota b; T = 20 °C. * The stability constant K for binding of chloroquine to C2IIa channels is given for comparison and was taken from Bachmeyer et al. (2003) [45].

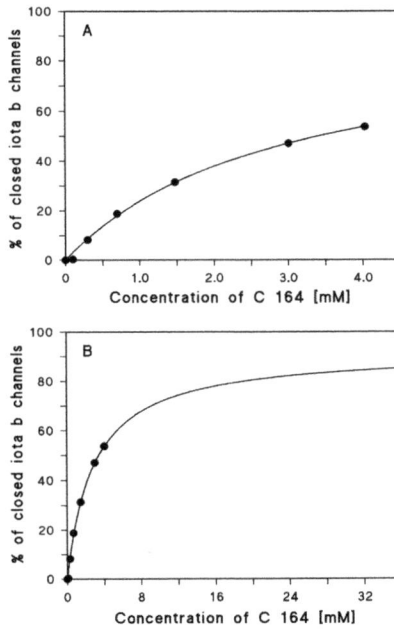

Figure 2. (**A**) Langmuir isotherm of the inhibition of Iota b-induced membrane conductance (about 70 Iota b-channels) by the aminoquinolinium salt C 164. The fit line corresponds to the data points taken from the titration experiment in Figure 1. The fit of the data was performed using Equation (2). The stability constant, K, for binding of C 164 to the Iota b-channels was (348 ± 48) $1/M$ (The channel block was at maximum 92% $\pm$ 7%; $K_S = 2.9$ mM ($r^2 = 0.997645$)); (**B**) Because of the low degree of inhibition in C 164 concentration range, we extrapolated its concentration to about 33 mM and used the same fit parameters as in A. The fit curve indicates that high concentration of C 164 almost fully blocked the Iota b channels.

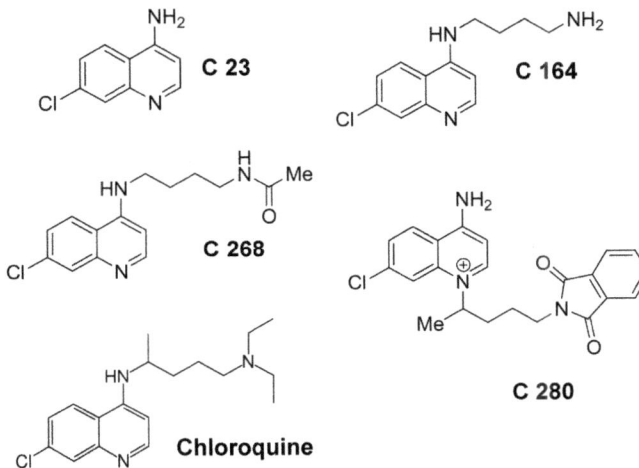

Figure 3. Structures of chloroquine and the chloroquine analogs (aminoquinolinium salts) used in this study). The chloroquine analogs were designated as suggested by Lödige (2013) [46].

2.2. C 280 Inhibited pH-Dependent Membrane Translocation of C2 and Iota Toxin

Prompted by the observation that C 280 interfered with the C2IIa and Iota b pores in vitro, we addressed the question whether C 280 also inhibits translocation of the enzyme components C2I and Iota a through their respective toxin pores across the membranes of intact cells. To this end, we used an established assay, which mimics the acidic conditions of endosomes on the surface of intact cells and allows direct translocation of the C2I and Iota a enzyme components into the cytosol through CIIa and Iota b pores, respectively, which were inserted in the plasma membrane under acidic conditions [34]. All steps of this assay were performed in the presence of Baf A1 to block the normal uptake of C2 and Iota toxins into the cytosol via acidified endosomes.

Vero cells were exposed to either C2 or Iota toxin under acidic conditions in the presence and absence of C 280 and subsequently the cells were further incubated under neutral conditions also with or without C 280. As shown in Figure 4, C2- and Iota-treated cells rounded up consequently to the acidic pulse. However, significantly less cells were rounded in the presence of C 280, indicating that less C2I or Iota a reached the cytosol. This result strongly suggests that C 280 inhibits membrane translocation of C2I and Iota a through the lumen of the C2IIa and Iota b pores, respectively. This result suggested that chloroquine-analogs similar to the structure of C 280 might provide interesting tools for the further study of aminoquinolinium salts as blockers for intoxication by C2-toxin. The block of intoxication of Iota-toxin by the same molecules does not look so promising because of the possible different structure of the Iota b channel (see Discussion).

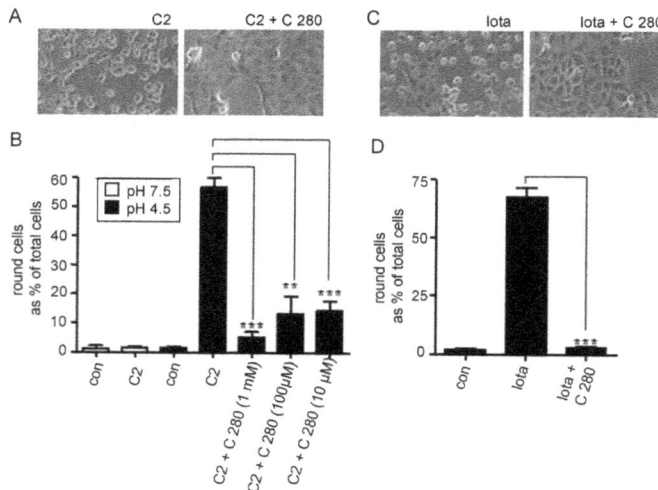

Figure 4. Inhibitory effect of C 280 on the pH-dependent trans-membrane transport of the C2 and iota toxins in living cells. (**A**) Baf A1-treated Vero cells were incubated for 30 min at 4 °C with C2 toxin (100 ng/mL C2I + 200 ng/mL C2IIa) to enable toxin binding. Noteworthy, Baf A1 was present to inhibit in a later step the "normal" transport of the A components of the internalized toxins into the cytosol via acidified endosomes, which is a prerequisite to investigate the toxin transport across the cytoplasmic membrane in this approach. For control (con), cells were incubated without toxin. Subsequently, cells were exposed for 5 min at 37 °C to acidic medium (pH 4.5 or to neutral medium pH 7.5 for control) and subsequently incubated at 37 °C in neutral medium containing Baf A1. During the acidic pulse, the B components insert as pores into the cytoplasmic membrane and the A components translocate through these pores into the cytosol of the cells and induce ADP-ribosylation of actin and cell-rounding. In this approach, the toxin-induced cell-rounding serves as an established specific and sensitive endpoint to monitor the uptake of the A components into the cytosol in the presence and absence of the inhibitor.

To test the effect of C 280 on toxin translocation, the indicated concentrations of C 280 were present in the medium during acidic pulse and the subsequent incubation periods. Pictures were taken after 1 and 2 h to document cell rounding, i.e., intoxication of cells (shown in A for 2 h and 1 mM C 280); (**B**) The percentage of intoxicated cells was determined after 1 h, and values are given as mean $\pm$ S.D. ($n = 3$). Significance was tested between cells, which have been treated with C2 toxin either in the absence or presence of C 280 by using the student's *t*-test (*** $p < 0.0005$, ** $p < 0.005$); (**C**) To test the influence of C 280 on membrane translocation of iota toxin, Baf A1-treated Vero cells were exposed for 15 min at ph 4.0 to Iota toxin (100 ng/mL Ia + 200 ng/mL Ib) in the presence or absence of 1 mM C 280. Cells were incubated for further 2 h at 37 °C in neutral medium containing Baf A1 in the presence or absence of C 280. Pictures were taken and the percentage of round cells was determined (**D**) Values are given as mean $\pm$ S.D. ($n = 3$). Significance was tested between cells, which have been treated with iota toxin either in the absence or presence of C 280 by using the student's t-test (*** $p < 0.0005$, ** $p < 0.005$).

3. Discussion

3.1. The Structure of the Aminoquinolinium Salts Allows an Interesting Insight in Structural Elements Required for Efficient Binding Protein Channel Blocking

The experiments with different aminoquinolinium salts to block the C2IIa channel allow an interesting insight in the structural requirement for efficient channel blocking, which can be considered as important tool for further in vivo and in vitro studies. The simplest chloroquine analog in the study here is C 23 (4-amino-7-chloroquinoline) which represents only the heterocyclic (bicyclic) part of the chloroquine molecule without side chain. C 23 has the smallest binding affinity to the C2IIa channels with a half saturation constant of about 700 µM for binding to C2IIa. The addition of n-butylamine to the amino group at the bicyclic molecule C 23 decreased the half saturation constant K_S for binding to C2IIa by a factor of more than 10 to 54 µM. The further addition of an acetyl group to the amino group of C 164 led to an additional decrease of the half saturation constant for the resulting C 268 molecule to about 5 µM. The half saturation constant is already below that of chloroquine, which means that C 268 is already a more efficient blocker of C2IIa channels than chloroquine. Another big step to improve binding of the aminoquinolinium salts to the C2IIa channel is the attachment of the bulky side chain to the nitrogen in the pyridine ring of C 23 (see Figure 3). The resulting chloroquine analog (C 280) has the highest affinity for binding to C2IIa but also for binding to Iota b. The half saturation constant for C 280 binding to C2IIa drops down to 0.16 µM, which is about 60-times lower than the half saturation constant for chloroquine binding.

The situation is in the case of binding of the aminoquinolinium salts to the Iota b channel not such straightforward as in the case of C2IIa. The reason for this is that the half saturation constant for binding shows some increase from C 23 to C 164 and then starts to become smaller in the series C 164, C 268, and C 280 (see Table 1). However, all half saturation constants for binding of the aminoquinolinium salts to Iota b were considerably higher than those for their binding to C2IIa. Even C 280, which had the highest binding affinity to Iota b with a half saturation constant of 80 µM showed a half saturation constant that was about 500 times higher than the corresponding constant for binding to C2IIa, which is remarkable. Similarly, the half saturation constant for binding of C 280 to the Iota b channel (80 µM) is only little smaller that for binding of chloroquine to the channel (140 µM), which looks a little strange when the situation is again compared to that of C2IIa. Nevertheless, the binding of C 280 to the Iota b channel is still strong enough that inhibition of the translocation of Iota a through Iota b is blocked in cell-based assays. This means that also certain aminoquinolinium salts could serve as blockers for intoxication of cells by Iota-toxin.

3.2. What Could Be the Reason That Blockers of Channel Function Have a Much Smaller Affinity to Iota b Than to C2IIa?

The results presented here and in previous studies demonstrated that chloroquine had much lower binding affinity to Iota b channels as compared to binding to C2IIa heptamers [37–40]. Similarly, the affinity of the aminoquinolinium salts to both binding protein channels differed considerably (see Table 1). The reason for this discrepancy is not quite clear because many structural elements that probably contribute to the binding site in the vestibule of the two channels are present in both primary sequences. There are the two rings of—at maximum—seven negatively charged residues in the binding site of the heptamer (D386 in Iota b corresponding to E399 in C2IIa and D413 in Iota b corresponding to D426 in C2IIa [41]). The only difference is that Iota b contains aspartate in position 386 whereas C2IIa has in the corresponding position 399 a glutamate. However, this very conservative exchange should not interfere much with the properties of the binding site in the vestibule of the two heptamers. Similarly, both cannels contain an Φ-clamp (F415 in Iota b and F428 in C2IIa), which is also an important structural element in both channels [40,41]. Thus, it is only slightly understandable that they differ so substantially in binding of the aminoquinolinium salts. The only remarkable difference between Iota b and C2IIa is the number of negatively charged groups within the channel-forming domain itself [41]. Whereas the membrane-spanning beta-sheet structure of C2IIa contains one glutamate (E307), there are no charges in the membrane spanning part of Iota b. This could represent the difference. However, experiments with a C2IIa mutant where E307 was replaced by lysine demonstrated that the E307K C2IIa mutant channel had approximately the same affinity for chloroquine as wildtype C2IIa [47]. This means presumably that charges within the pore-forming domain of the binding protein heptamers are most likely not essential for aminoquinolinium salt binding. This has to do with the strong image force along the channel that is created by the many negatively charged groups within the vestibule of the channels [47]. Taken together, it seems moreover that the structure of the binding site within the vestibule of the Iota b heptamers shows some structural differences to those of C2IIa and PA that result in a lower affinity for binding of chloroquine and its analogs [41,48].

3.3. The Aminoquinolinium Salts Inhibit the Trans-Membrane Transport of the A Components of the Binary C2 and Iota Toxins Through the Pores Formed under Acidic Conditions by the B Components in Membranes of Living Cells and Protect Cells from Intoxication with These Toxins

As expected from the in vitro results with lipid bilayers, the compounds inhibited the pH-triggered trans-membrane transport of the enzyme components of C2-toxin and Iota-toxin through the pores formed by the B-components of these toxins in cell membranes under acidic conditions, thereby protecting the cells from intoxication. This was tested in a well-established experimental approach where the situation of acidic endosomal vesicles was experimentally mimicked on the surface of living cultured epithelial cells. Vero cells were pre-treated with Baf A1 to inhibit endosomal acidification and thereby prevent the "normal" uptake of the binary C2 and iota toxins via acidic endosomes [17–19]. Then, the cells were incubated for 30 min at 4 °C with the respective toxin to enable toxin binding to the cell surface receptors but not receptor-mediated endocytosis, which does not occur at 4 °C. Subsequently, the cells were exposed to a short acidic pulse at 37 °C to trigger the conformational change of the cell-bound B components, which then form pores in the cytoplasmic membrane and mediate the transport of their bound A components through these pores across the cytoplasmic membrane into the cytosol of the cells. This transport of the A components into the cytosol results in ADP-ribosylation of actin and cell-rounding, which serves as a highly specific and sensitive endpoint to monitor the uptake of the A components into the cytosol in the presence and absence of the inhibitor. No cell-rounding was observed without toxin, indicating that the acidic conditions alone had no effect on cell morphology, or when cells were exposed to neutral medium because in this case, Baf A1 prevented the toxin uptake into the cytosol (Figure 4). As shown in Figure 4, cells only rounded up when the toxin was bound prior to the acidic pulse, indicating translocation of the A components into the cytosol. Significantly fewer cells rounded up under such conditions in the presence of C 280, clearly indicating the inhibitory effect of this compound on the membrane transport of the binary toxins C2 and iota in living cells. The aminoquinolinium salts had no effect on cell morphology under the experimental conditions used in this study. They have also the unique advantage that their toxicology is known from malaria treatment. In conclusion, these compounds should represent attractive lead compounds for development of novel pharmacological inhibitors against binary clostridial actin ADP-ribosylating toxins and may further be used against related binary toxins from pathogenic bacteria.

4. Experimental Procedures

4.1. Materials

The recombinant components of C2 toxin, C2I and C2II, were expressed as GST fusion proteins in *Escherichia (E.) coli* BL21 cells and purified as described [8]. To obtain biologically active C2IIa, C2II was treated with trypsin as reported earlier [8]. Iota a and Iota b were kind gifts of Dr. Michel R. Popoff (Institut Pasteur, Paris, France) [11]. Iota a and Iota b were activated by α-chymotrypsin as described previously [11,40,49]. The heterocyclic chloroquine analogs (aminoquinolinium salts) C 23, C 164, C 268, and C 280 (see Figure 3) were kind gifts of Dr. Gerhard Bringmann and Dr. Melanie Lödige, Institute for Organic Chemistry, University of Würzburg, 97074 Würzburg, Germany. The chloroquine analogs were termed and synthesized as was described in detail recently [46]. The chloroquine analogs were dissolved in ultrapure water supplemented with 10% (*v/v*) ETOH. Cell culture media (DMEM, MEM) and fetal calf serum were obtained from Invitrogen (Karlsruhe, Germany) and cell culture materials from Techno Plastic Products. (Trasadingen, Switzerland). Complete® protease inhibitor and streptavidin-peroxidase were from Roche (Mannheim, Germany), Baf A1 from Calbiochem (Bad Soden, Germany), and biotinylated NAD+ from R&D Systems GmbH (Wiesbaden-Nordenstadt, Germany).

All salts (analytical grade) were obtained from Sigma-Aldrich Chemie GmbH (München, Germany) and were dissolved in ultrapure H_2O (Milli-Q® systems, Merck Millipore, Darmstadt, Germany). Diphytanoyl phosphatidylcholine (DiPhPC) was obtained from Avanti Polar Lipids Alabaster AL and n-decane (analytical grade from Merck, Darmstadt, Germany).

4.2. Methods

4.2.1. Cell Culture and Cytotoxicity Tests

African green monkey kidney (Vero) cells were cultivated at 37 °C and 5% CO_2 in MEM containing 10% FCS, 1.5 g/L sodium bicarbonate, 1 mM sodium-pyruvate, 2 mM L-glutamine, 0.1 mM non-essential amino acids. Vero cells were reseeded twice a week for, at most, 15–20 times. The macrophage-like murine J774A.1 cells were cultivated at 37 °C and 5% CO_2 in DMEM containing 10% FCS and 4 mM L-glutamate. For cytotoxicity experiments with C2-toxin or Iota-toxin, Vero cells were incubated at 37 °C in 1 mL serum-free medium containing both components of C2 toxin (200 ng/mL C2IIa + 100 ng/mL C2I) or iota toxin (200 ng/mL Iota b + 100 ng/mL Iota a). After different incubation periods, the toxin-induced cell-rounding was documented with a Zeiss Axiovert 40CFl microscope (Zeiss, Oberkochen, Germany) containing a Jenoptik progress C10 CCD camera (Carl Zeiss GmbH, Jena, Germany) and the percentage of round cells was determined from the pictures [19]. Inhibitory effects of C 280 were analyzed by incubating the cells with toxin in the presence of C 280.

4.2.2. Toxin-translocation Assay with Intact Vero Cells

The pH-dependent translocation of C2 toxin across the cytoplasmic membranes of intact Vero cells was performed as described earlier [8]. In brief, Vero cells were pretreated for 30 min at 37 °C with Baf A1 (100 nM) to prevent normal internalization of the toxin via acidified endosomes. Subsequently, cells were incubated at 4 °C in serum-free medium with C2IIa (200 ng/mL) and C2I (100 ng/mL) to enable toxin binding. Cells were washed and exposed for 5 min to warm acidic medium (37 °C, pH 4.5, Baf A1) to trigger insertion of cell-bound C2IIa into the cytoplasmic membrane and subsequent translocation of C2I through the pores across the membrane. Subsequently, the cells were further incubated at 37 °C in complete medium under neutral conditions in the presence of Baf A1 and C2I-toxin induced cell-rounding was documented by photography. The pH-driven translocation of cell-bound Iota toxin across the cytoplasmic membrane of Vero cells was performed as described earlier by Blöcker et al. (2001) [50]. Baf A1-treated cells were exposed for 15 min at 37 °C to acidic medium (pH 4.0) containing Iota toxin (100 ng/mL Iota a + 200 ng/mL Iota b) and subsequently incubated at 37 °C in neutral medium containing Baf A1. The number of round cells was determined to document the cytopathic action of Iota toxin. To test an inhibitory effect of C 280 on membrane translocation of C2- and Iota-toxins, C 280 was applied to the medium during the acidic pulse and the subsequent incubation periods, pictures from the cells were taken and the number of round cells was determined from the pictures.

4.2.3. Lipid Bilayer Experiments

The experiments with planar lipid bilayers were performed as has been described previously in detail [51]. In brief membranes were formed by the painting method using DiPhPC dissolved to 1% (*w/v*) in n-decane. The membrane hole had an area of about 0.4 mm^2 in the thin wall separating two 5 mL compartments in a Teflon cell. The different binding components (C2IIa and Iota b) were added from concentrated solutions to the aqueous phase either immediately before membrane formation or after the membranes had turned black in concentrations of about 1 to 10 ng/mL. The temperature was maintained at 20 °C during all experiments. The membrane conductance induced by channels formed by the binding components C2IIa and Iota b was measured after application of fixed membrane potentials with a pair of silver/silver chloride electrodes with salt bridges inserted into the aqueous compartments on both sides of the DiPhPC membranes. The electrodes were connected in series to a voltage source and a homemade current-to-voltage converter made with a Burr Brown operational amplifier. The amplified signal was monitored on a digital storage oscilloscope (OWON) and recorded on a strip chart recorder.

4.2.4. Titration Experiments with the Different Aminoquinolinium Salts

The binding of the compounds C 23, C 164, C 268, and C 280 (see Figure 3) to the channels formed by the binding components was investigated with titration experiments similar to those used previously to study the binding of carbohydrates to the LamB-channel of Escherichia coli [52] and the binding of tailored azolopyridinium salts to channels formed by protective antigen (PA) [33] and C2IIa [34]. About 30 min after start of the reconstitution of the binding components into lipid bilayer membranes, their reconstitution rate in the membranes became very small. Then concentrated solutions of different aminoquinolinium salts were added to both sides of the membranes while stirring to allow equilibration. The results of the titration experiments were analyzed in a similar way as has been performed previously for the binding of azolopyridinium salts to channels formed by protective antigen (PA) [33] and C2IIa [34]. The conductance, $G(c)$ of the membrane at a given concentration, c, of the different aminoquinolinium salts C 23, C 164, C 268, and C 280 relative to the initial conductance, $Gmax$ (in the absence of the ligands), was analyzed using the following equation, which corresponds to Langmuir adsorption isotherms [33,34]:

$$\% \text{ of fraction of blocked channels} = \frac{(G_{\max} - G(c))}{G_{\max}} = \frac{100 \cdot K \cdot c}{(K \cdot c + 1)} \tag{1}$$

K is the stability constant for the binding of aminoquinolinium salts to channels formed by Iota b and C2IIa [33]. The half saturation constant, K_S of this process is given by the inverse stability constant $1/K$. K can be derived from the titration experiments by a fit of the experimental data to Equation (1). We did not observe full channel blockage in all titration experiments. In cases of only partial blockage of the C2IIa and Iota b channels, Equation (1) had to be modified to account for the reduced maximum blockage given by A in percent (the maximum degree A of blockage was in all cases between 85% to 100%):

$$\% \text{ of fraction of blocked channels} = \frac{(G_{\max} - G(c))}{G_{\max}} = \frac{K \cdot c}{(K \cdot c + 1)} \tag{2}$$

Acknowledgments: The authors would like to thank Gerhard Bringmann and Melanie Lödige, Institute for Organic Chemistry, University of Würzburg for the kind gift of the aminoquinolinium salts C 23, C 164, C 268, and C 280 and for their contribution at the early stage of this work. We thank Michel R. Popoff, Institut Pasteur Paris, France for providing the Iota-toxin. We thank Ulrike Binder, University of Ulm, and Jennifer Rausch, University of Würzburg for expert technical assistance. This work was financially supported by the Deutsche Forschungsgemeinschaft (DFG, grant BA 2087/2-2).

Author Contributions: Angelika Kronhardt, Christoph Beitzinger, Holger Barth, and Roland Benz conceived and designed the experiments; Angelika Kronhardt, Christoph Beitzinger, and Holger Barth performed the experiments; Holger Barth and Roland Benz analyzed the data; Holger Barth and Roland Benz wrote the paper.

Conflicts of Interest: The authors declare no conflict of interest.

References

1. Barth, H.; Aktories, K.; Popoff, M.R.; Stiles, B.G. Binary bacterial toxins: Biochemistry, biology, and applications of common Clostridium and Bacillus proteins. *Microbiol. Mol. Biol. Rev.* **2004**, *68*, 373–402. [CrossRef] [PubMed]

2. Aktories, K.; Barth, H. New insights into the mode of action of the actin ADP-ribosylating virulence factors *Salmonella enterica* SpvB and *Clostridium botulinum* C2 toxin. *Eur. J. Cell Biol.* **2011**, *90*, 944–950.

3. Knapp, O.; Benz, R.; Popoff, M.R. Pore-forming activity of clostridial binary toxins. *Biochim. Biophys. Acta* **2016**, *1858*, 512–525. [CrossRef] [PubMed]

4. Friedlander, A.M. Macrophages are sensitive to anthrax lethal toxin through an acid-dependent process. *J. Biol. Chem.* **1986**, *261*, 7123–7126. [PubMed]

5. Mock, M.; Fouet, A. Anthrax. *Annu. Rev. Microbiol.* **2001**, *55*, 647–671. [CrossRef] [PubMed]

6. Collier, R.J.; Young, J.A. Anthrax toxin. *Annu. Rev. Cell. Dev. Biol.* **2003**, *19*, 45–70. [CrossRef] [PubMed]

7. Ohishi, I.; Iwasaki, M.; Sakaguchi, G. Purification and characterization of two components of botulinum C2 toxin. *Infect. Immun.* **1980**, *30*, 668–673. [PubMed]

8. Barth, H.; Blöcker, D.; Behlke, J.; Bergsma-Schutter, W.; Brisson, A.; Benz, R.; Aktories, K. Cellular uptake of *Clostridium botulinum* C2 toxin requires oligomerization and acidification. *J. Biol. Chem.* **2000**, *275*, 18704–18711. [CrossRef] [PubMed]

9. Stiles, B.G.; Wilkins, T.D. Purification and characterization of *Clostridium perfringens* iota toxin: Dependence on two nonlinked proteins for biological activity. *Infect. Immun.* **1986**, *54*, 683–688. [PubMed]

10. Popoff, M.R.; Boquet, P. Clostridium spiroforme toxin is a binary toxin which ADP-ribosylates cellular actin. *Biochem. Biophys. Res. Commun.* **1988**, *152*, 1361–1368. [CrossRef]

11. Gibert, M.; Petit, L.; Raffestin, S.; Okabe, A.; Popoff, M.R. *Clostridium perfringens* iota-toxin requires activation of both binding and enzymatic components for cytopathic activity. *Infect Immun.* **2000**, *68*, 3848–3853. [CrossRef] [PubMed]

12. Popoff, M.R.; Stiles, B.G. Bacterial toxins and virulence factors targeting the actin cytoskeleton and intercellular junctions. In *The Comprehensive Sourcebook of Bacterial Toxins*; Alouf, J.E., Popoff, M.R., Eds.; Academic Press: London, UK, 2006.

13. Schleberger, C.; Hochmann, H.; Barth, H.; Aktories, K.; Schulz, G.E. Structure and action of the binary C2 toxin from *Clostridium botulinum*. *J. Mol. Biol.* **2006**, *364*, 705–715. [CrossRef] [PubMed]

14. Petosa, C.; Collier, R.J.; Klimpel, K.R.; Leppla, S.H.; Liddington, R.C. Crystal structure of the anthrax toxin protective antigen. *Nature* **1997**, *385*, 833–838. [CrossRef] [PubMed]

15. Nguyen, T.L. Three-dimensional model of the pore form of anthrax protective antigen. Structure and biological implications. *J. Biomol. Struct. Dyn.* **2004**, *22*, 253–265. [CrossRef] [PubMed]

16. Jiang, J.; Pentelute, B.L.; Collier, R.J.; Zhou, Z.H. Atomic structure of anthrax protective antigen pore elucidates toxin translocation. *Nature* **2015**, *521*, 545–549. [CrossRef] [PubMed]

17. Haug, G.; Leemhuis, J.; Tiemann, D.; Meyer, D.K.; Aktories, K.; Barth, H. The host cell chaperone Hsp90 is essential for translocation of the binary *Clostridium botulinum* C2 toxin into the cytosol. *J. Biol. Chem.* **2003**, *278*, 32266–32274. [CrossRef] [PubMed]

18. Kaiser, E.; Pust, S.; Kroll, C.; Barth, H. Cyclophilin A facilitates translocation of the *Clostridium botulinum* C2 toxin across membranes of acidified endosomes into the cytosol of mammalian cells. *Cell. Microbiol.* **2009**, *11*, 780–795. [CrossRef] [PubMed]

19. Kaiser, E.; Kroll, C.; Ernst, K.; Schwan, C.; Popoff, M.; Fischer, G.; Buchner, J.; Aktories, K.; Barth, H. Membrane translocation of binary actin-ADP-ribosylating toxins from *Clostridium difficile* and *Clostridium perfringens* is facilitated by cyclophilin A and Hsp90. *Infect. Immun.* **2011**, *79*, 3913–3921. [CrossRef] [PubMed]

20. Zhang, S.; Finkelstein, A.; Collier, R.J. Evidence that translocation of anthrax toxin's lethal factor is initiated by entry of its N terminus into the protective antigen channel. *Proc. Natl. Acad. Sci. USA* **2004**, *101*, 16756–16761. [CrossRef] [PubMed]

21. Young, J.A.; Collier, R.J. Anthrax toxin: Receptor binding, internalization, pore formation, and translocation. *Annu. Rev. Biochem.* **2007**, *76*, 243–265. [CrossRef] [PubMed]

22. Aktories, K.; Bärmann, M.; Ohishi, I.; Tsuyama, S.; Jakobs, K.H.; Habermann, E. Botulinum C2 toxin ADP-ribosylates actin. *Nature* **1986**, *322*, 390–392. [CrossRef] [PubMed]

23. Vandekerckhove, J.; Schering, B.; Bärmann, M.; Aktories, K. Botulinum C2 toxin ADP-ribosylates cytoplasmic beta/gamma-actin in arginine 177. *J. Biol. Chem.* **1988**, *263*, 696–700. [PubMed]

24. Schering, B.; Barmann, M.; Chhatwal, G.S.; Geipel, U.; Aktories, K. ADP-ribosylation of skeletal muscle and non-muscle actin by *Clostridium perfringens* iota toxin. *Eur. J. Biochem.* **1988**, *171*, 225–229. [CrossRef] [PubMed]

25. Heine, K.; Pust, S.; Enzenmüller, S.; Barth, H. ADP-ribosylation of actin by *Clostridium botulinum* C2 toxin in mammalian cells results in delayed caspase-dependent apoptotic cell death. *Infect. Immun.* **2008**, *76*, 4600–4608. [CrossRef] [PubMed]

26. Popoff, M.R.; Bouvet, P. Clostridial toxins. *Future Microbiol.* **2009**, *4*, 1021–1064. [CrossRef] [PubMed]

27. Gülke, I.; Pfeifer, G.; Liese, J.; Fritz, M.; Hofmann, F.; Aktories, K.; Barth, H. Characterization of the enzymatic component of the ADP-ribosyltransferase toxin CDTa from *Clostridium difficile*. *Infect. Immun.* **2001**, *69*, 6004–6011. [CrossRef] [PubMed]

28. Perelle, S.; Gibert, M.; Bourlioux, P.; Corthier, G.; Popoff, M.R. Production of a complete binary toxin (actin-specific ADP-ribosyltransferase) by *Clostridium difficile* CD196. *Infect. Immun.* **1997**, *65*, 1402–1407. [PubMed]

29. Popoff, M.R.; Rubin, E.J.; Gill, D.M.; Boquet, P. Actin-specific ADP-ribosyltransferase produced by a *Clostridium difficile* strain. *Infect. Immun.* **1988**, *56*, 2299–2306. [PubMed]

30. Popoff, M.R.; Boquet, P. *Clostridium spiroforme* toxin is a binary toxin which ADP-ribosylates cellular actin. *Biochem. Biophys. Res. Commun.* **1988**, *152*, 1361–1368. [CrossRef]

31. Han, S.; Craig, J.A.; Putnam, C.D.; Carozzi, N.B.; Tainer, J.A. Evolution and mechanism from structures of an ADP-ribosylating toxin and NAD complex. *Nat. Struct. Biol.* **1999**, *6*, 932–936. [PubMed]

32. Leuber, M.; Orlik, F.; Schiffler, B.; Sickmann, A.; Benz, R. Vegetative insecticidal protein (Vip1Ac) of *Bacillus thuringiensis* HD201: Evidence for oligomer and channel formation. *Biochemistry* **2006**, *45*, 283–288. [CrossRef] [PubMed]

33. Beitzinger, C.; Bronnhuber, A.; Duscha, K.; Riedl, Z.; Huber-Lang, M.; Benz, R.; Hajós, G.; Barth, H. Designed azolopyridinium salts block protective antigen pores in vitro and protect cells from anthrax toxin. *PLoS ONE* **2013**, *8*. [CrossRef] [PubMed]

34. Bronnhuber, A.; Maier, E.; Riedl, Z.; Hajós, G.; Benz, R.; Barth, H. Inhibitions of the translocation pore of *Clostridium botulinum* C2 toxin by tailored azolopyridinium salts protects human cells from intoxication. *Toxicology* **2014**, *316*, 25–33. [CrossRef] [PubMed]

35. Nestorovich, E.M.; Karginov, V.A.; Popoff, M.R.; Bezrukov, S.M.; Barth, H. Tailored ß-cyclodextrin blocks the translocation pores of binary exotoxins from *C. botulinum* and *C. perfringens* and protects cells from intoxication. *PLoS ONE* **2011**, *6*. [CrossRef] [PubMed]

36. Roeder, M.; Nestorovich, E.M.; Karginov, V.A.; Schwan, C.; Aktories, K.; Barth, H. Tailored cyclodextrin pore blocker protects mammalian cells from *Clostridium difficile* binary toxin CDT. *Toxins (Basel)* **2014**, *6*, 2097–2114. [CrossRef] [PubMed]

37. Schmid, A.; Benz, R.; Just, I.; Aktories, K. Interaction of *Clostridium botulinum* C2 toxin with lipid bilayer membranes. Formation of cation-selective channels and inhibition of channel function by chloroquine. *J. Biol. Chem.* **1994**, *269*, 16706–16711. [PubMed]

38. Bachmeyer, C.; Benz, R.; Barth, H.; Aktories, K.; Gilbert, M.; Popoff, M.R. Interaction of *Clostridium botulinum* C2 toxin with lipid bilayer membranes and Vero cells: Inhibition of channel function by chloroquine and related compounds in vitro and intoxification in vivo. *FASEB J.* **2001**, *15*, 1658–1660. [CrossRef] [PubMed]

39. Knapp, O.; Benz, R.; Gibert, M.; Marvaud, J.C.; Popoff, M.R. Interaction of *Clostridium perfringens* iota-toxin with lipid bilayer membranes. Demonstration of channel formation by the activated binding component Ib and channel block by the enzyme component Ia. *J. Biol. Chem.* **2002**, *277*, 6143–6152. [CrossRef] [PubMed]

40. Knapp, O.; Maier, E.; Waltenberger, E.; Mazuet, C.; Benz, R.; Popoff, M.R. Residues involved in the pore-forming activity of the *Clostridium perfringens* iota toxin. *Cell Microbiol.* **2015**, *17*, 288–302. [CrossRef] [PubMed]

41. Neumeyer, T.; Schiffler, B.; Maier, E.; Lang, A.E.; Aktories, K.; Benz, R. *Clostridium botulinum* C2 toxin. Identification of the binding site for chloroquine and related compounds and influence of the binding site on properties of the C2II channel. *J. Biol. Chem.* **2008**, *283*, 3904–3914. [CrossRef] [PubMed]

42. Krantz, B.A.; Melnyk, R.A.; Zhang, S.; Juris, S.J.; Lacy, D.B.; Wu, Z.; Finkelstein, A.; Collier, R.J. A phenylalanine clamp catalyzes protein translocation through the anthrax toxin pore. *Science* **2005**, *309*, 777–781. [CrossRef] [PubMed]

43. Melnyk, R.A.; Collier, R.J. A loop network within the anthrax toxin pore positions the phenylalanine clamp in an active conformation. *Proc. Natl. Acad. Sci. USA* **2006**, *103*, 9802–9807. [CrossRef] [PubMed]

44. Song, L.; Hobaugh, M.R.; Shustak, C.; Cheley, S.; Bayley, H.; Gouaux, J.E. Structure of staphylococcal alpha-hemolysin, a heptameric transmembrane pore. *Science* **1996**, *274*, 1859–1866. [CrossRef] [PubMed]

45. Bachmeyer, C.; Orlik, F.; Barth, H.; Aktories, K.; Benz, R. Mechanism of C2-toxin inhibition by fluphenazine and related compounds: Investigation of their binding kinetics to the C2II-channel using the current noise analysis. *J. Mol. Biol.* **2003**, *333*, 527–540. [CrossRef] [PubMed]

46. Lödige, M. Synthese und Evaluierung Neuartiger Wirkstoffklassen Gegen Infektionskrankheiten (Synthesis and Evaluation of Novel Drug Classes Against Infectious Diseases). Ph.D. Thesis, University of Würzburg, Würzburg, Germany, 29 July 2013.

47. Blöcker, D.; Bachmeyer, C.; Benz, R.; Aktories, K.; Barth, H. Channel formation by the binding component of *Clostridium botulinum* C2 toxin: Glutamate 307 of C2II affects channel properties in vitro and pH-dependent C2I translocation in vivo. *Biochemistry* **2003**, *42*, 5368–5377. [CrossRef] [PubMed]

48. Orlik, F.; Schiffler, B.; Benz, R. Anthrax toxin protective antigen: Inhibition of channel function by chloroquine and related compounds and study of binding kinetics using the current noise analysis. *Biophys. J.* **2005**, *88*, 1715–1724. [CrossRef] [PubMed]

49. Ohishi, I. Activation of botulinum C2 toxin by trypsin. *Infect. Immun.* **1987**, *55*, 1461–465. [PubMed]

50. Blöcker, D.; Behlke, J.; Aktories, K.; Barth, H. Cellular uptake of the binary *Clostridium perfringens* iota toxin. *Infect. Immun.* **2001**, *69*, 2980–2987. [CrossRef] [PubMed]

51. Benz, R.; Janko, K.; Boos, W.; Lauger, P. Formation of large, ion-permeable membrane channels by the matrix protein (porin) of *Escherichia coli*. *Biochim. Biophys. Acta* **1978**, *511*, 305–319. [CrossRef]

52. Benz, R.; Schmid, A.; Vos-Scheperkeuter, G.H. Mechanism of sugar transport through the sugar-specific LamB channel of *Escherichia coli* outer membrane. *J. Membr. Biol.* **1987**, *100*, 21–29. [CrossRef] [PubMed]

MDPI AG

St. Alban-Anlage 66

4052 Basel, Switzerland

Tel. +41 61 683 77 34

Fax +41 61 302 89 18

http://www.mdpi.com

Toxins Editorial Office

E-mail: toxins@mdpi.com

http://www.mdpi.com/journal/toxins

www.ingramcontent.com/pod-product-compliance
Lightning Source LLC
Chambersburg PA
CBHW051912210326
41597CB00033B/6125